ENVIRONMENTAL RESEARCH ADVANCES

THE REGIONAL GREENHOUSE GAS INITIATIVE

LESSONS AND RELATED CAP-AND-TRADE CONSIDERATIONS

Environmental Research Advances

Additional books in this series can be found on Nova's website under the Series tab.

Additional E-books in this series can be found on Nova's website under the E-book tab.

ENVIRONMENTAL RESEARCH ADVANCES

THE REGIONAL GREENHOUSE GAS INITIATIVE

LESSONS AND RELATED CAP-AND-TRADE CONSIDERATIONS

SERGE CHAUVIN
EDITOR

NOVINKA

New York

For permission to use material from this book please contact us:
Telephone 631-231-7269; Fax 631-231-8175
Web Site: http://www.novapublishers.com

NOTICE TO THE READER

Additional color graphics may be available in the e-book version of this book.

LIBRARY OF CONGRESS CATALOGING-IN-PUBLICATION DATA

ISBN: 978-1-62808-483-2

Published by Nova Science Publishers, Inc. † New York

CONTENTS

PREFACE

This book provides an overview of the RGGI cap-and-trade program and the participating RGGI states, with a discussion on select issues raised by RGGI that may be of interest to policymakers who are considering developing a federal program.

Chapter 1 – The Regional Greenhouse Gas Initiative (RGGI) is the nation's first mandatory cap-and-trade program for greenhouse gas (GHG) emissions. As of January 1, 2012, RGGI involves nine states—Connecticut, Delaware, Maine, Maryland, Massachusetts, New Hampshire, New York, Rhode Island, and Vermont. The RGGI cap-and-trade system applies only to carbon dioxide (CO_2) emissions from electric power plants with capacities to generate 25 megawatts or more— approximately 168 facilities. The RGGI emissions cap took effect January 1, 2009.

The initial results of the RGGI program may be instructive to policymakers. Several of RGGI's design elements generated considerable interest during the development and debate of federal proposals to address GHG emissions. In particular, the program's emission cap has received particular attention. Since the cap took effect in 2009, it has not compelled regulated entities to make internal emission reductions or purchase emission credits from other sources. Several factors led to this outcome: RGGI's cap design, an economic downturn, and a substantial shift to less carbon intensive fuels. In 2005, RGGI states generated 32% of their electricity from coal and petroleum, sources of energy with relatively high carbon intensity. In 2011, these sources generated 12% of RGGI's electricity. To address the disparity between the cap and actual emissions, in February 2013, RGGI participants proposed to substantially reduce the existing cap. The proposed cap would take effect in 2014 and would be based on 2012 emission levels. However,

RGGI state legislatures and/or agencies must alter relevant statutes and/or regulations before the proposed cap can take effect.

Although RGGI's emission cap has had limited impact on the region's power plant emissions, the program has had other effects. The cap's existence (coupled with unlimited emission allowance banking and an auction reserve price) attaches a price to the regulated entities' CO_2 emissions. Because the cap is currently non-binding, this price acts like an emissions fee or carbon tax.

RGGI states have sold 89% of their emission allowances through quarterly auctions. The auction proceeds—over $1.2 billion to date—have provided a new source of state revenue, which have been used to support various policy objectives. RGGI states (as a group) have contributed the majority of the emission allowance value (65%) to support energy efficiency, renewable energy, or other climate-related efforts. Several RGGI studies indicate that supporting energy efficiency provides multiple benefits: emission reduction, consumer savings via lower electricity bills, and job creation. However, RGGI states have demonstrated that revenue allocation strategies are subject to change. For example, after initially allotting auction proceeds to energy efficiency efforts, three states have transferred auction proceeds to address state budget deficits, drawing criticism from some environmental groups.

As a group, the nine RGGI states account for approximately 7% of U.S. CO_2 emissions (and 16% of U.S. Gross Domestic Product). RGGI's aggregate emissions rank in the top 20 among all nations. But from a practical standpoint, the RGGI program's contribution to directly reducing the global accumulation of GHG emissions in the atmosphere is arguably negligible. However, RGGI's activities may stimulate action in other states or at the federal level: when confronted with a growing patchwork of state/regional requirements, industry stakeholders may support a singular national policy.

Chapter 2 – When designing a cap-and-trade program, one of the more controversial and challenging questions for policymakers is *how, to whom, and for what purpose* to distribute the emission allowances. Regarding the *method* of distribution to covered sources, policymakers could (1) sell the allowances through an auction process, (2) allocate the allowances at no cost to covered sources, (3) provide allowances to non-covered sources who would, in turn, sell them to covered sources, or (4) use some combination of these methods. Although the emission allocation method *would not* affect the environmental integrity of the cap-and-trade program, the selected allocation strategy could have considerable consequences.

Using auctions as a distribution method could avoid certain concerns that are likely to occur if covered sources receive all (or most) of the allowances at no cost: (1) consumers in different electricity markets may face inequitable price increases; (2) a weak price signal for electricity may be sent in areas with the most carbon-intensive fuel portfolios; and (3) no-cost allowances may overcompensate covered sources. In addition, auction revenues offer a unique opportunity to reduce the overall costs of the emissions program. Several economic studies indicate that if used in the most efficient manner, overall costs could be minimized by almost 50%.

A greenhouse gas (GHG) emission cap-and-trade program would create a valuable new commodity: the GHG emission allowance. EPA estimates that allowance value could potentially account — in aggregate — for tens or hundreds of billions of dollars each year. When distributing this value, policymakers would face a choice between minimizing the costs imposed on the entire economy, minimizing the expected burden on specific parties, or supporting a range of climate- or nonclimate-related policy objectives.

For example, Congress may consider providing transition assistance to carbon-intensive industries. Studies have estimated profits could be maintained in the energy production and electricity generation sectors, if approximately 20% of allowances were provided to those sectors at no cost. Members may also consider allotting allowance value to consumers, particularly low-income households, who are expected to bear the majority of the compliance costs via higher energy prices. Another option would involve distributing the allowance value to support various objectives: technology development, energy efficiency improvements, biological sequestration, climate change adaptation efforts, or non climate-related purposes, such as deficit reduction. Of these objectives, technology advancement is arguably the most crucial in terms of mitigation. Moreover, deployment of new technologies could potentially lower the overall costs of the program.

Although many of the proposals in the 110th Congress (e.g., S. 2191, S. 1766, and S. 3036) would employ an auction to some degree, none of the bills specifies the design of the auction. Congress may want to consider including specific design elements in legislative text, particularly auction frequency and whether or not the auction should have a reserve price, and if so, at what level.

Chapter 3 – If Congress establishes a greenhouse gas (GHG) emissions reduction program (e.g., cap-and-trade system), the treatment of GHG emission offsets would likely be a critical design element. If allowed as part of an emissions program, offsets could provide cost savings and other benefits. However, offsets have generated concern.

An offset is a measurable reduction, avoidance, or sequestration of GHG emissions from a source not covered by an emission reduction program. If allowed, offset projects could generate "emission credits," which could be used by a regulated entity (e.g., power plant) to comply with its reduction requirement. Offsets could include various activities:

- agriculture or forestry projects: e.g., conservation tillage or planting trees on previously non-forested lands;
- renewable energy projects: e.g., wind farms;
- energy efficiency projects: e.g., equipment upgrades;
- non-CO_2 emissions reduction projects: e.g., methane from landfills.

Including offsets would likely make an emissions program more cost-effective by (1) providing an incentive for non-regulated sources to generate emission reductions and (2) expanding emission compliance opportunities for regulated entities. Some offset projects may provide other benefits, such as improvements in air or water quality. In addition, the offset market may create new economic opportunities and spur innovation as parties seek new methods of generating offsets.

The main concern with offset projects is whether or not they represent real emission reductions. For offsets to be credible, a ton of CO_2-equivalent emissions from an offset project should equate to a ton reduced from a covered emission source, such as a smokestack or exhaust pipe. This objective presents challenges because many offsets are difficult to measure. If illegitimate offset credits flow into an emissions trading program, the program would fail to reduce GHG emissions. Another concern is whether the inclusion of offsets would send the appropriate price signal to encourage the development of long-term mitigation technologies. Policymakers may consider a balance between price signal and program costs.

If eligible in a U.S. program, international offsets are expected to dominate in early decades because they would likely offer the lowest-cost options. Domestic sectors, such as agriculture and forestry, might benefit if international offsets are excluded. Some object to the use of international offsets due to concerns of fairness: the low-cost options would be unavailable to developing nations if and when they establish GHG emission targets. However, some offset projects may promote sustainable development. On the other hand, international offsets may serve as a disincentive for developing nations to enact laws or regulations controlling GHG emissions because many projects would no longer qualify as offsets.

In: The Regional Greenhouse Gas Initiative
Editor: Serge Chauvin ISBN: 978-1-62808-483-2

Chapter 1

THE REGIONAL GREENHOUSE GAS INITIATIVE: LESSONS LEARNED AND ISSUES FOR POLICYMAKERS*

Jonathan L. Ramseur

SUMMARY

The Regional Greenhouse Gas Initiative (RGGI) is the nation's first mandatory cap-and-trade program for greenhouse gas (GHG) emissions. As of January 1, 2012, RGGI involves nine states—Connecticut, Delaware, Maine, Maryland, Massachusetts, New Hampshire, New York, Rhode Island, and Vermont. The RGGI cap-and-trade system applies only to carbon dioxide (CO_2) emissions from electric power plants with capacities to generate 25 megawatts or more— approximately 168 facilities. The RGGI emissions cap took effect January 1, 2009.

The initial results of the RGGI program may be instructive to policymakers. Several of RGGI's design elements generated considerable interest during the development and debate of federal proposals to address GHG emissions. In particular, the program's emission cap has received particular attention. Since the cap took effect in 2009, it has not compelled regulated entities to make internal emission reductions or purchase emission credits from other sources. Several factors led to this

---

* This is an edited, reformatted and augmented version of a Congressional Research Service publication, CRS Report for Congress R41836, prepared for Members and Committees of Congress, from www.crs.gov, dated May 21, 2013.

outcome: RGGI's cap design, an economic downturn, and a substantial shift to less carbon intensive fuels. In 2005, RGGI states generated 32% of their electricity from coal and petroleum, sources of energy with relatively high carbon intensity. In 2011, these sources generated 12% of RGGI's electricity. To address the disparity between the cap and actual emissions, in February 2013, RGGI participants proposed to substantially reduce the existing cap. The proposed cap would take effect in 2014 and would be based on 2012 emission levels. However, RGGI state legislatures and/or agencies must alter relevant statutes and/or regulations before the proposed cap can take effect.

Although RGGI's emission cap has had limited impact on the region's power plant emissions, the program has had other effects. The cap's existence (coupled with unlimited emission allowance banking and an auction reserve price) attaches a price to the regulated entities' CO_2 emissions. Because the cap is currently non-binding, this price acts like an emissions fee or carbon tax.

RGGI states have sold 89% of their emission allowances through quarterly auctions. The auction proceeds—over $1.2 billion to date—have provided a new source of state revenue, which have been used to support various policy objectives. RGGI states (as a group) have contributed the majority of the emission allowance value (65%) to support energy efficiency, renewable energy, or other climate-related efforts. Several RGGI studies indicate that supporting energy efficiency provides multiple benefits: emission reduction, consumer savings via lower electricity bills, and job creation. However, RGGI states have demonstrated that revenue allocation strategies are subject to change. For example, after initially allotting auction proceeds to energy efficiency efforts, three states have transferred auction proceeds to address state budget deficits, drawing criticism from some environmental groups.

As a group, the nine RGGI states account for approximately 7% of U.S. CO_2 emissions (and 16% of U.S. Gross Domestic Product). RGGI's aggregate emissions rank in the top 20 among all nations. But from a practical standpoint, the RGGI program's contribution to directly reducing the global accumulation of GHG emissions in the atmosphere is arguably negligible. However, RGGI's activities may stimulate action in other states or at the federal level: when confronted with a growing patchwork of state/regional requirements, industry stakeholders may support a singular national policy.

Introduction

A number of states and local governments have taken action to directly address greenhouse gas (GHG) emissions. These efforts cover a wide spectrum, from developing climate action plans to setting mandatory GHG emission standards.[1]

One of the most significant climate change developments at the state level is the Regional Greenhouse Gas Initiative (RGGI, pronounced "Reggie"), the nation's first mandatory cap-andtrade program (see text box, "What Is a Cap-and-Trade System?") for GHG emissions, which went into effect January 1, 2009.[2] RGGI involves nine states—Connecticut, Delaware, Maine, Maryland,[3] Massachusetts,[4] New Hampshire, New York, Rhode Island, and Vermont.[5]

Several of RGGI's design elements generated considerable interest during the development and debate of federal proposals to address GHG emissions. The initial results of the RGGI program may be instructive to policymakers, because RGGI may serve as a possible test case for a federal cap-and-trade program, providing insights into implementation complexities, the mechanics of various design elements, and lessons of potential design pitfalls.

The first section of this report provides an overview of the RGGI cap-and-trade program and the participating RGGI states. The second section discusses selected issues raised by RGGI that may be of interest to policymakers who are considering developing a federal program. The final section provides some final thoughts concerning the RGGI program.

What Is a Cap-and-Trade System?

A GHG cap-and-trade system creates an overall limit (i.e., a cap) on GHG emissions from the emission sources covered by the program. Cap-and-trade programs can vary by the sources covered. The covered sources, also referred to as covered or regulated entities, often include major emitting sectors (e.g., power plants and carbon-intensive industries), fuel producers/processors (e.g., coal mines or petroleum refineries), or some combination of both.

The emissions cap is partitioned into emission allowances. Typically, in a GHG cap-and-trade system, one emission allowance represents the authority to emit one (metric) ton of carbon dioxide-equivalent (tCO_2-e). This term of measure is used because GHGs vary by global warming potential (GWP)—an index of how much a GHG may contribute to global warming over a period of time, typically 100 years. GWPs are used to compare gases to CO_2, which has a GWP of 1. For example, methane's GWP is 25, because a ton of methane is 25 times more potent a GHG than a ton of CO_2.

Under an emissions cap, covered entities with relatively low emission-reduction costs have a financial incentive to make reductions beyond what is required, because these further reductions could be sold (i.e., traded) as emission credits to entities that face higher costs to reduce their facility emissions. At the end of each established compliance period (e.g., a calendar year or multiple years), covered sources surrender emission allowances (to an implementing agency) to cover the number of tons emitted. If a source did not provide enough allowances to cover its emissions, the source would be subject to penalties. Other mechanisms, such as banking or offsets, may be included to increase the flexibility of the program.

The emissions cap creates a new currency—the emission allowance. Policymakers may decide to distribute the emission allowances to covered entities at no cost (based on, for example, previous years' emissions), sell the allowances (e.g., through an auction), or use some combination of these strategies. The distribution of emission allowance value is typically a source of intense debate during a cap-and-trade's program development.

A cap-and-trade program is one policy tool for reducing GHG emissions. It is often described as a market-based mechanism, because it (like an emissions fee or carbon tax) allows the marketplace to determine the economically efficient solution for GHG emission reduction. Compared to more traditional approaches—requiring, for example, specific performance standards or technologies at particular facilities—market-based mechanisms are generally considered more cost effective. Perhaps the most successful market-based program in the environmental policy arena is the sulfur dioxide (SO_2) emissions trading system (also called the Acid Rain Program) established by the 1990 amendments to the Clean Air Act.

RGGI OVERVIEW

RGGI is sector-specific cap-and-trade system that only applies to carbon dioxide (CO_2) emissions from electric power plants[6] in RGGI states with capacities to generate 25 megawatts or more[7]— approximately 168 facilities.[8] RGGI designers expected the initial program to be a foundation for emissions trading and possibly expanded in future years by covering other emission sources/sectors, GHGs, or other states. The existing program addresses approximately 22% of all GHG emissions in RGGI states.[9]

Compared to recent federal proposals, the reductions required by RGGI's cap are relatively modest. As the cap indicates (see *Figure 1* and its surrounding discussion), the initial objective of RGGI was to stabilize CO_2 emissions for several years (2009-2014) at the expected 2009 levels (based on assumptions made in 2005), and then require gradual reductions, achieving a 10% decrease from the 2009 emission cap level by 2019. As discussed below, RGGI states have proposed to modify the existing cap and make additional changes to particular design elements.

RGGI's cap-and-trade program includes many of the design elements that have been proposed and debated in federal legislative proposals. Highlights include the following:

- *Three-year compliance periods.* At the end of a compliance period covered entities submit one emission allowance for each ton of CO_2 emissions generated. A three-year compliance period should mitigate potential emission allowance price swings brought on by short-term market volatility.
- *Emission allowance banking.* RGGI allows covered entities to bank an unlimited number of emission allowances for future use. The opportunity to bank emission allowances instills a substantial amount of flexibility into a trading program, effectively making annual emissions caps flexible over time, and reduces the absolute cost of compliance.
- *Offsets use.* An offset is a measurable reduction, avoidance, or sequestration of GHG emissions from a source not covered by an emission reduction program. To a limited degree (discussed below), covered entities may submit offsets in lieu of the emission allowances needed to satisfy compliance obligations.
- *Emission allowance auctions.* A substantial percentage (89%) of emission allowances have been distributed through quarterly auctions (discussed below). The auctions include a reserve price, which sets a price floor for emission allowances.
- *Consumer benefit allocation.* RGGI states agreed that at least 25% of the emission allowance value would be distributed "for a consumer benefit or strategic energy purpose."[10] Allowance value distributions from RGGI states have exceeded this minimum requirement (discussed below).
- *Safety-Valve.*[11] RGGI provides an additional year to demonstrate compliance if emission allowance prices reach a certain level. In

addition, covered entities may cover a greater proportion of their emissions with offsets. However, RGGI states proposed to eliminate these features and replace them with a "cost containment reserve" (CCR) in 2014. The CCR would provide additional allowances—5 million in 2014 and 10 million each year thereafter—if certain price thresholds were met: $4/ton in 2014; $6/ton in 2015; $8/ton in 2016; and $10/ton in 2017, increasing 2.5% each year thereafter. Unlike other allowance reserve proposals,[12] allowances from CCR would not be borrowed from future years, thus effectively increasing the cap if triggered.

SELECTED ISSUES

The following selected RGGI issues may be of interest to policymakers seeking to develop a federal program that would reduce greenhouse gas emissions.

Emissions Cap

Although RGGI is one of the more aggressive state programs addressing climate change, the program's current emission cap has exceeded actual emissions since its inception (*Figure 1*). This means that since the cap took effect in 2009, the cap has not *compelled* regulated entities to make internal emission reductions or purchase emission credits (or offsets). As discussed below, the degree to which this has occurred was unexpected.

In February 2013, RGGI participants proposed to substantially reduce the existing cap.[13] However, RGGI state legislatures and/or agencies must alter relevant statutes and/or regulations before the proposed cap can take effect. This section includes a discussion of both the existing cap and the proposed cap.

Existing Cap

During RGGI's construction in 2005,[14] RGGI designers set the 2009 emissions cap about 4% above the average emission levels observed between 2000 and 2002. RGGI designers anticipated that power plant CO_2 emissions would gradually increase, so that actual levels would approximately match the cap set for 2009—188 million short tons of CO_2 ($mtCO_2$). Actual emissions did not meet these projections, but decreased substantially (*Figure 1*).

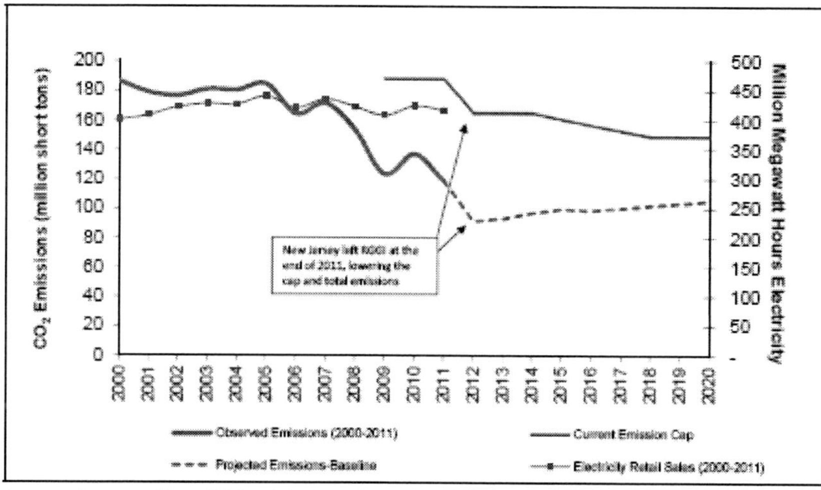

Source: Prepared by CRS with data from the following: observed state emission data (2000-2011) provided by RGGI, at http://www.rggi.org/market/tracking/public_reporting; projected RGGI emissions from modeling results, available at http://www.rggi.org/design/program_review; electricity retail sales data from Energy Information Administration, at http://www.eia.gov/electricity/data.cfm.

Notes: The "Projected Emissions-Baseline" estimate assumes that the cap remains unchanged. Additional assumptions are available at http://www.rggi.org/design/program_review/materials-by-topic/modeling.

Figure 1. Observed and Projected RGGI Power Plant CO_2 Emissions Compared to RGGI Emissions Cap.

Studies that examined RGGI emissions and their underlying factors project that the region's CO_2 emissions will not return to 2005 levels in the near term, but will remain below the existing cap at least through 2020 (*Figure 1*).[15] These studies indicate that RGGI CO_2 emission decreases (2005- 2011) were due (to some degree) to long-term, structural changes, such as changes in RGGI's electricity generation portfolio and energy efficiency improvements. A comparison between the emission decline and electricity use in the RGGI states supports this notion. As *Figure 1* indicates, RGGI electricity retail sales (a proxy for electricity use) decreased by 5% between 2005 and 2011, while CO_2 emissions from in-state electricity generation decreased by 36%. This disparity suggests that factors others than temporal economic conditions are the primary influence.

Figure 2 compares RGGI's electricity generation portfolio between 2005 and 2011. The figure depicts a substantial decline in carbon intensive electricity generation over that timeframe. Electricity is generated from a variety of energy sources, which vary significantly by their ratio of CO_2 emissions per unit of energy. For example, a coal-fired power plant emits almost twice as much CO_2 (per unit of energy) as a natural gas-fired facility.[16] Some energy sources (e.g., hydropower, nuclear, wind, or solar) are generally considered to be zero-emission sources. In 2005, RGGI states generated 32% of their electricity from coal and petroleum, sources of energy with relatively high carbon intensity. In 2011, these sources generated 12% of RGGI's electricity. During that time, sources with relatively lower carbon intensities—natural gas, nuclear, and hydroelectric—increased their contributions to RGGI's electricity portfolio.

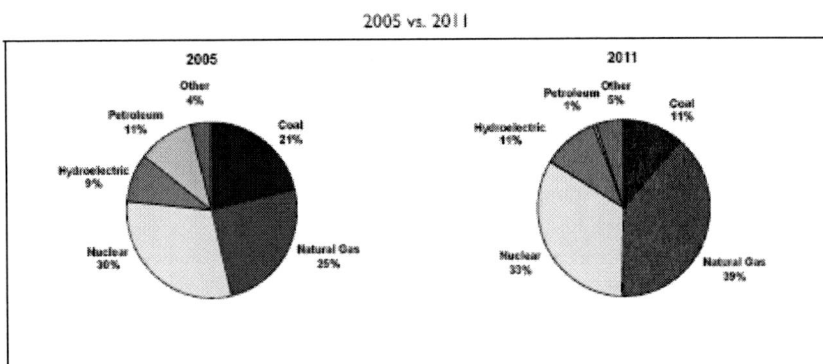

Source;: Prepared by CRS; data from Energy Information Administration, Electric Power Annual, historical data tables, at http://www.eia.gov/cneaf/electricity/epa/epa_sum.html.

Note: The "other" category includes wood and other biomass, wind, solar, and other gases.

Figure 2. RGGI Electricity Generation by Energy Source.

Proposed Emissions Cap

As part of its 2012 design review of the RGGI program,[17] participating states proposed to substantially reduce the emissions cap. Starting in 2014, the cap would decrease from 165 $mtCO_2$ to 91 $mtCO_2$. RGGI designers chose this level because it is expected to match 2012 emissions. Between 2015 and 2020, the cap will decline by 2.5% each year. In addition, RGGI states proposed to lower the cap further to account for the substantial amount of banked emission allowances held by RGGI entities.[18] RGGI designers estimate that banked emissions—purchased between 2009 and 2013—will total 115 $mtCO_2$. Thus, the cap adjustments are considerable. In some years, the adjustments lower the cap by 20 $mtCO_2$, an almost 25% decrease.

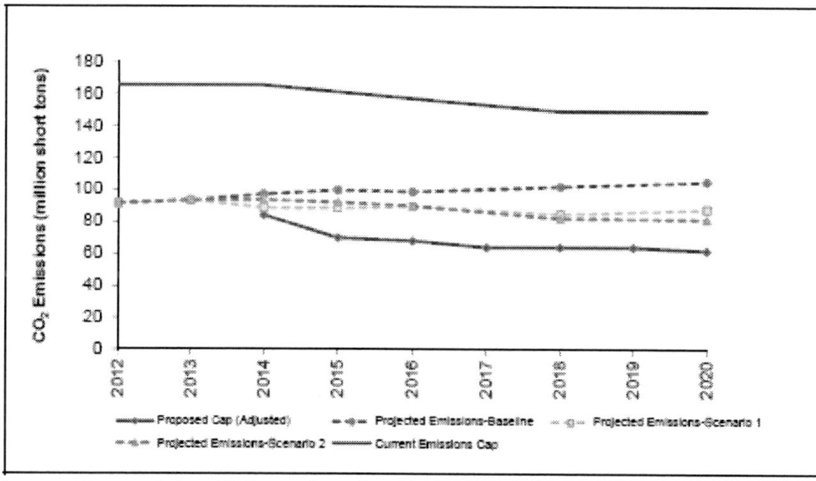

Source: Prepared by CRS; projected emission data from RGGI modeling results, available at http://www.rggi.org/design/program_review.

Notes: The proposed cap (adjusted) is based on RGGI model estimates of the number of emission allowances banked between 2009 and 2013 (115 $mtCO_2$). The baseline emission projection assumes that the existing emissions cap and other design elements remain unchanged. The other emission scenarios assume the proposed adjusted cap and other new elements, including the Cost Containment Reserve, take effect in 2014. The estimates from Scenario 1 and Scenario 2 are different because of varied assumptions in the underlying model. For further details about these estimates and the underlying assumptions, see http://www.rggi.org/design/ program_review.

Figure 3. RGGI Existing and Proposed Emission Caps and Projected Emissions.

Figure 3 illustrates (1) the adjusted cap, based on the RGGI estimate of banked emission allowances, (2) the existing emissions cap, and (3) three projections of CO_2 emissions from RGGI power plants. The figure indicates that under "baseline" conditions (i.e., the existing cap and other design elements remain intact), emissions would increase by approximately 15% between 2012 and 2020. If the proposed emissions cap and other design changes are finalized by the states, RGGI models estimate ("Scenario 1" and "Scenario 2" in *Figure 3*) that emissions in 2020 would be 3%-10% lower than in 2012. The range of reductions reflects different assumptions made in the two projection scenarios. *Figure 3* indicates that in both emission projection scenarios, the 2020 emissions are above the proposed cap.

Emission Cap Impacts

Although RGGI's existing emission cap has not required emission reductions since its inception, the program still has impacts. First, the cap's existence attaches a price to the regulated entities' CO_2 emissions. The price is relatively low (as discussed below), because of the abundance of emission allowances. A 2010 analysis of the RGGI program found that the emission allowance price accounted for approximately 3.4% of the change in the price difference between natural gas and coal in the RGGI region between 2005 and 2009.[19]

Second, the cap's emission allowances are essentially a new form of currency. Their value can be used to support various policy objectives, including (as is the case with RGGI) energy efficiency and renewable energy investments. Some would argue that RGGI's greatest impact so far is to provide a relatively reliable funding source for such efforts. Several RGGI studies indicate that supporting energy efficiency provides multiple benefits: emission reduction, consumer savings via lower electricity bills, and regional job creation.[20] One of these studies argues this allowance value distribution strategy (e.g., use of RGGI auction revenue) "creates wider benefit than any other use of allowance value."21 As discussed below, such efforts play a role in determining the effectiveness of the program.

Emissions Leakage

A critical design detail—electricity imports from non-RGGI states—remains unresolved. This presents an opportunity for "emissions leakage," which could undermine the effectiveness of the RGGI program. Leakage can occur when an emissions reduction program does not include all sources contributing to the environmental problem. Increases in emissions from uncovered sources may reduce emission achievements from covered sources.

This is a concern with the RGGI program, because the RGGI regime does not regulate emissions from electricity generated outside the region (e.g., in Pennsylvania) and then used within the region (i.e., "imported electricity"). As illustrated by *Figure 4,* over the past two decades the RGGI states (as a group) have imported between 5% and 11% of their annual electricity needs (measured in electricity sales). After several RGGI milestones—the December 2005 Memorandum of Agreement and the January 2009 start date of the emissions cap—the imported electricity has remained within this range. However, as discussed above the cap (so far) has had minimal effect on the relative price of different sources of electricity. The proposed emissions cap may increase emission allowance prices and increase the possibility of emissions leakage, but other factors (e.g., the relative price of natural gas) may counter this potential impact.

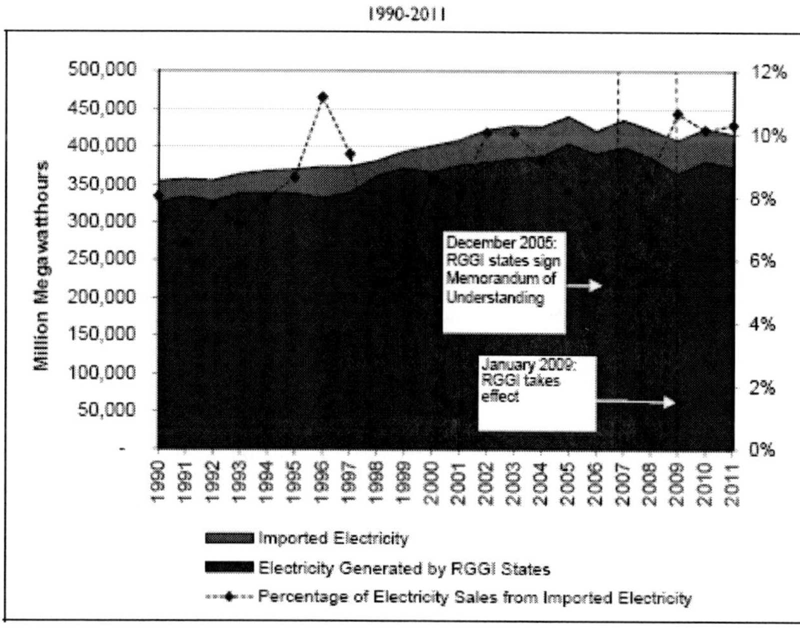

Source: Prepared by CRS; data from Energy Information Administration.

Notes: Electricity imports calculated by subtracting total electricity generation from total electricity sales. Although the emissions cap became effective January 1, 2009, RGGI held its first emission allowance auction in September 2008.

Figure 4. Electricity Sales, Generation, and Imports in the RGGI States.

Emissions leakage can occur if imported electricity replaces RGGI in-state electricity generation, because emissions from in-state electricity are regulated; emissions from imported electricity are not. In such a scenario, the quantity of leakage would depend on the sources of electricity generation involved in the tradeoff. For example, maximum leakage would occur if imported electricity from a coal-fired power plant replaced in-state electricity generated from a zero-emission source.

A RGGI working group issued a final report on leakage March 2008. Among the four recommendations, the working group concluded that states should monitor for emissions leakage and evaluate whether more direct measures should be considered at a later date.[22] After its 2012 Program Review, RGGI participants seemed to reach a similar conclusion, stating

> the states commit, over the course of the next year, to engage in a collaborative effort ... to identify and evaluate potential imports tracking tools, conduct further modeling to ascertain energy and price implications of any potential policy on emissions associated with imported electricity, and pursue additional legal research necessary, leading to a workable, practicable, and legal mechanism to address emissions associated with imported electricity.[23]

Emission Allowance Value Distribution

When designing a cap-and-trade program, one of the more controversial and challenging questions for policymakers is how, to whom, and for what purpose to distribute the emission allowances. Individual RGGI states determine how their allowances are distributed. RGGI states have answered the "how question" by employing auctions to distribute the vast majority of allowances. During the first compliance period (2009-2011), states' percentages of auctioned allowances ranged from 57% to 99%, but as a group, the states auctioned 89% of all available allowances.[24] The remaining portion of allowances are distributed to various entities to support a variety of objectives. These distributions vary by state and are subject to change.[25] Perhaps the more important question for policymakers is what to do with the emission allowance value—in the case of RGGI, allowance value predominately means auction revenues.

Allowance Auctions

RGGI's auctions may be of particular interest to Congress, because this approach was part of several proposed cap-and-trade systems in recent years.[26] Each RGGI auction is conducted in one round with a sealed-bid, uniform price format.[27] Participants may submit multiple, confidential bids for a certain number of allowances at a specific price. The price paid by all bidders is the highest rejected bid (i.e., the second-highest bid). For example, consider a hypothetical auction, in which the supply of allowances is 20 units. The highest bidder offers $10 per allowance for 15 allowances.

The second highest bidder offered $9 per allowance for 10 allowances. Under RGGI's auction structure, the highest bidder would receive 15 allowances at $9/allowance, and the second-highest bidder would receive 5 allowances at $9/allowance. The price paid by all successful bidders is known as the clearing price.

In addition, RGGI auctions include a reserve price, below which the seller refuses to part with the item for sale (i.e., emission allowance). The reserve price started at $1.86 in 2009, increasing to $1.98 in 2013.[28]

In a large volume, multi-unit auction that is expected to have substantial participation (i.e., high demand for emission allowances), a reserve price would all but guarantee a revenue stream. A reserve price may address certain logistical concerns, such as bidder collusion, that may be associated with auctions. In addition, a reserve price may provide assurance to parties making emission reductions that the reductions will have a minimal value in the allowance market.

Because the RGGI emissions cap has exceeded and is projected to exceed emissions (until the cap is adjusted in 2014), the reserve price acts like an emissions fee or carbon tax.[29] As discussed below, the revenue raised by the auctions may lead to emission reduction by supporting energy efficiency and renewable energy projects.

The RGGI program has held 19 auctions as of the date of this report. In general, many have viewed the auctions as successful in terms of price discovery,[30] transparency, transaction costs, and other logistical issues.

Another typical measure of auction success is revenue generation. After 19 auctions, the cumulative proceeds total over $1.2 billion. *Figure 5* illustrates the auctions' results. As the figure depicts, the clearing price equaled the reserve price in auctions conducted between June 2010 and December 2012, reflecting the abundance of emission allowances in the market. Moreover, during this time period, approximately 40% of the allowances offered for sale were not purchased. RGGI states retired the vast majority of these unsold allowances.[31]

In the most recent auction (March 2013), the clearing price exceeded the reserve price. This is likely related to the February 2013 proposal to substantially reduce the emissions cap in 2014.

September 2008 – March 2013

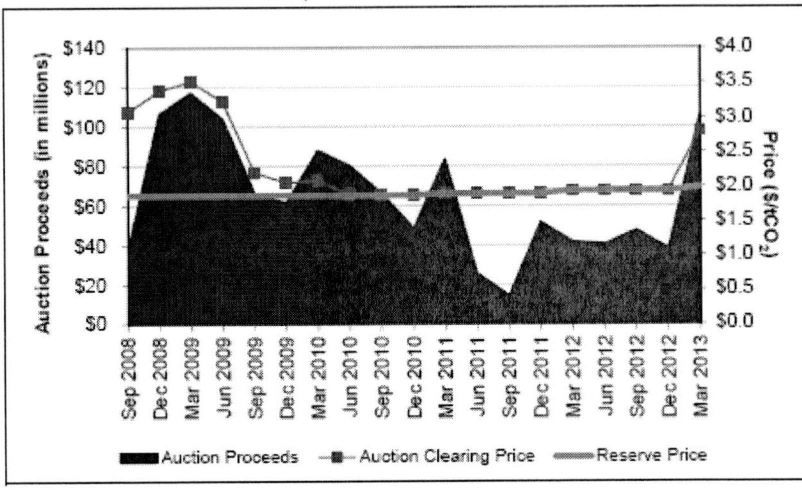

Source: Prepared by CRS; data from RGGI Inc., at http://www.rggi.org/.
Note: The reserve price has risen from $1.86 in 2008 to $1.98 in 2013.

Figure 5. RGGI Auctions: Proceeds and Clearing Prices.

Emission Allowance Value Distribution

When deciding how to distribute the emission allowance value, policymakers face trade-offs that could have considerable consequences.[32] In both RGGI's 2005 Memorandum of Understanding and subsequent Model Rule,[33] states agreed that at least 25% of emission allowance value would be allocated for a "consumer benefit or strategic energy purpose."[34] As of September 2012, RGGI states (as a group) have more than doubled this minimum commitment. RGGI states have allocated more than 65% of the emission allowance value to support energy efficiency, renewable energy, and other emission reduction efforts.[35]

Allowance value includes revenues generated through allowance auctions or by giving the allowances away at no charge to either covered or non-covered entities. A covered entity recipient could use the allowances for compliance purposes, sell the allowances in the marketplace, or bank the allowances for future use. To realize the value of allowances received, a non-covered entity recipient would need to sell the allowances in the marketplace, either through a broker or directly to a covered entity.

Table 1 details emission allowance value distribution by state. Of the various objectives, RGGI states (as a group) have contributed the most—61% of allowance value—to support energy efficiency and renewable energy, but the allotments vary considerably by state (*Table 1*). Moreover, RGGI states have demonstrated that allowance value distribution decisions are subject to change. For example, after initially allotting auction proceeds to energy efficiency efforts, several states transferred auction proceeds to address state budget deficits. Environmental groups criticized the actions of these states, but the state policymakers argued that the transfers were necessary.

Table 1. Distribution of Emission Allowance Value by State Includes Auction Revenues and Set-Aside Allowance Value (as of September 2012)

State	Energy Efficiency and/or Renewable Energy	Other Climate Change-Related Efforts	Program Administration	State Budget Deficit Reduction	Free Allocation to Covered Entities	Allocation for Exemptions	Long-Term Power Agreements	Other Purposes
Connecticut	92%	0%	7%	0%	0%	0%	1%	0%
Delaware	46%	6%	6%	0%	27%	16%	0%	0%
Maine	95%	1%	4%	0%	0%	0%	0%	0%
Maryland	43%	8%	3%	0%	0%	9%	4%	32%
Massachusetts	97%	0%	2%	0%	0%	0%	0%	1%
New Hampshire	63%	0%	1%	6%	0%	0%	0%	30%
New Jersey	34%	7%	4%	46%	0%	1%	8%	0%
New York	56%	4%	11%	21%	0%	3%	2%	2%
Rhode Island	95%	0%	5%	0%	0%	0%	0%	0%
Vermont	98%	0%	2%	0%	0%	0%	0%	0%
RGGI Total	61%	4%	6%	13%	1%	3%	3%	9%

Source: Prepared by CRS; data from Environment Northeast, RGGI Auction Tracker: State Allocations and Spending Plans, September 2012. Notes: New Jersey left the program at the end of 2011. The New Jersey data are included for comparison purposes. The above table includes broad categories created by CRS. The Environment Northeast report provides a more detailed breakdown of emission allowance value distribution.

These developments highlight a cap-and-trade design issue for federal policymakers: how much flexibility (if any) should be built into a strategy to distribute emission allowance value. For example, should crafters include provisions that authorize modifying (without legislation) an enacted distribution approach, perhaps based on specific criteria?

OFFSETS

Federal policymakers may be interested in RGGI's treatment of offsets. An offset is a measurable reduction, avoidance, or sequestration of GHG emissions from a source not covered by an emission reduction program.[36] RGGI limits offsets to 3.3% of a source's allowance submission, a relatively low percentage compared to some federal proposals.[37] Under the current framework, RGGI increases this percentage to 5% or 10% if the market price of an allowance exceeds $7 or $10 (in 2005 dollars, adjusted annually), respectively. However, as part of the design review in February 2013, RGGI states proposed to eliminate this provision, due to the proposed Cost Containment Reserve.

RGGI uses a standards approach—as opposed to performance-based system—for developing offsets: projects must satisfy a set of detailed requirements (specific to a project type) and be certified by a third party. In contrast to recent federal proposals, which often allow a wide array of project types, RGGI limits offset projects to five types, which must be located in RGGI states:

- landfill methane reduction;
- sulfur hexafluoride reductions from specific industrial activities;
- planting trees where none were previously growing (afforestation);[38]
- specific energy efficiency projects;
- avoided methane from manure management practices.

Some offset projects raise concerns, because they may not represent real emission reductions. For offsets to be credible, a ton of CO_2-equivalent emissions from an offset project should equate to a ton reduced from a RGGI power plant. If illegitimate offset credits flow into an emissions trading program, the program would fail to achieve its primary goal—emission reduction. In general, the project types allowed in RGGI are considered to be of higher quality in terms of their ability to represent real, verifiable, and permanent emission reductions.[39]

According to the RGGI offsets tracking database, no offset projects have been developed under the RGGI program.[40] This is likely related to the low emission allowance price and the non-constraining emissions cap.

Interaction with Federal GHG Emission Regulations

As of the date of this report, EPA has proposed GHG emission regulations for new power plants but not existing power plants.[41] Environmental stakeholders are urging EPA to develop regulations that would apply to existing facilities, an action that would likely generate considerable interest and controversy among policymakers. An issue for both federal and state policymakers is how RGGI's emission program would interact with potential EPA regulations that apply to existing power plants.[42] For example, would RGGI requirements satisfy potential future obligations under federal rules?

FINAL THOUGHTS

As a group, the nine RGGI states account for approximately 7% of U.S. CO_2 emissions from energy consumption and 16% of the U.S. Gross Domestic Product (GDP).[43] *Table 2* indicates that RGGI's aggregate emissions rank in the top 20 among nations. But from a practical standpoint, the RGGI program's contribution to directly reducing the global accumulation of GHG emissions in the atmosphere is arguably negligible.

However, when business and industry have confronted a growing patchwork of state requirements, these sectors have historically preferred a national policy. RGGI and other state programs, particularly developments in California, may have some influence on federal policymakers. Note that the combination of RGGI and California CO_2 emissions would account for a substantial portion of U.S. CO_2 emissions (14%) and rank above Germany (*Table 2*).

In addition, RGGI's activities may create examples and/or models that will prove instructive for federal policymakers crafting more widespread applications. Moreover, the program has provided a training ground for personnel from multiple states and various professions to develop a specific expertise in emissions trading issues. This knowledge base would be useful if a federal system were developed.

Table 2. Top-Ranked Nations, U.S. States, and Selected Groups for CO_2 Emissions from Energy Consumption (2010 Data)

Country, State, or Group	CO_2 Emissions (million metric tons)	Country, State, or Group	CO_2 Emissions (million metric tons)
China	7,997	Canada	547
United States	5,637	United Kingdom	529
European Union	3,940	South Africa	473
Russian Federation	1,642	Saudi Arabia	469
India	1,601	Brazil	451
Japan	1,180	Mexico	432
Germany	793	Australia	424
Texas	653	RGGI (not including New Jersey)	418
South Korea	581	Italy	417
Iran	565	Indonesia	415

Source: Prepared by CRS with data from EIA, International Energy Statistics, Total CO_2 Emissions from the Consumption of Energy, at http://www.eia.gov.

Note: If New Jersey emissions from 2010 are included, the aggregate RGGI emissions are 553 mmtCO_2, approximately 10% of total U.S. CO_2 emissions.

Although the overall economic impacts directly related to RGGI's emissions cap have likely been minimal (because it is currently non-binding), differences among the RGGI states may lead to different economic impacts. The RGGI states differ—in some cases dramatically—by their percentages of total GHG emissions from CO_2 emissions from electricity generation (*Figure 6*). A key factor in these differences is the variance in states' carbon contents of electricity generation. As discussed earlier, electricity comes from an array of energy sources that vary significantly by their ratio of CO_2 emissions per unit of energy produced. States that generate a high percentage of electricity from coal compared to lower carbon alternatives will have a relatively high value for carbon content of electricity. For example, Vermont, which relies almost exclusively on nuclear, hydroelectricity, and renewables for electricity, has a very low carbon content of electricity generation, resulting in minimal CO_2 emissions from electricity generation. In contrast, other states (e.g., Delaware and Maryland) generate more of their electricity through coal combustion. Thus, these states' carbon contents of electricity are substantially higher, leading to higher percentages in *Figure 6.* When the emission cap becomes binding, the price of electricity is likely to increase at a sharper rate in these states. On the other hand, if a state invested in less carbon-intensive energy sources years before RGGI's development, energy consumers in that state may have already encountered economic impacts associated with the transition.

Compared to the entire United States, the nine RGGI states (in aggregate) generate a smaller portion of their GHG emissions from electricity generation (*Figure 6).* Moreover, none of the individual RGGI states exceeds the national figure of 40% (*Figure 6*). This difference is due to several factors. First, RGGI states (as a group) are net importers of electricity. Thus, some of the electricity consumed (about 10%, see *Figure 4*) in RGGI states is generated outside of the region. This reiterates the challenge in addressing GHG emissions from power plants at the state or regional level.

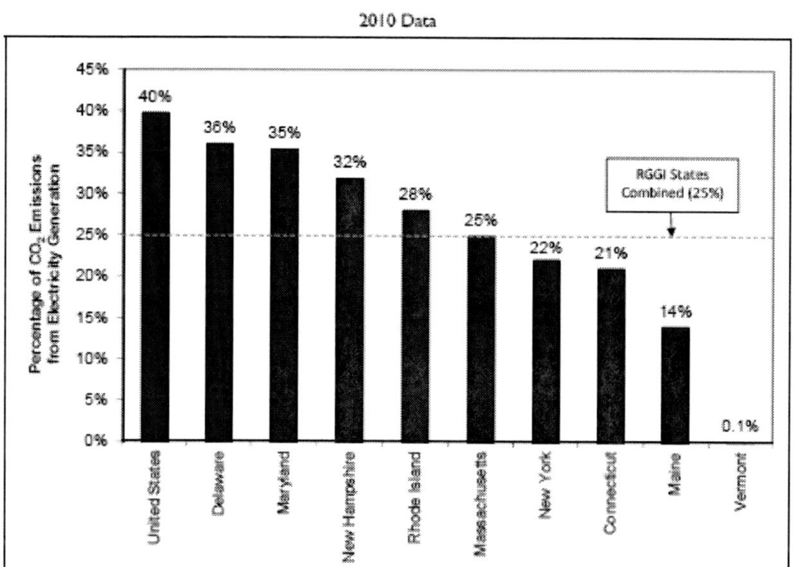

Source: Prepared by CRS; data from EIA, State CO_2 Emissions, at http://www.eia. gov/environment/emissions/ state/state_emissions.cfm.

Figure 6. Comparison of Percentages of CO_2 Emissions from Electricity Generation.

Second, the electricity generation profile of RGGI states (as a group) is less carbon-intensive than the overall United States.[44] Because RGGI states' electricity is relatively low in its carbon intensity, other sector sources of GHG emissions—transportation, industrial, residential, commercial—contribute a relatively larger portion of the region's overall GHG emissions compared to other parts of the country.

End Notes

[1] For example, California is implementing regulations that address GHG emissions on multiple fronts, including a capand-trade program that began in January 2013. For more information on the California cap-and-trade program, see the California Environmental Protection Agency and Air Resources Board website at http://www.arb.ca.gov/cc/capandtrade/capandtrade.htm.

[2] Prior to the starting date of the emissions cap (January 1, 2009), RGGI held its first emission allowance auction on September 25, 2008.

[3] Maryland Governor O'Malley signed RGGI's Memorandum of Understanding on April 20, 2007, making Maryland the first state that was not an original RGGI participant to join the regional initiative.

[4] Massachusetts and Rhode Island were involved in RGGI's development from the beginning. However, both states' governors declined to sign the Memorandum of Understanding in 2005. Massachusetts (different governor) and Rhode Island (same governor) joined RGGI as participants in January 2007.

[5] New Jersey participated in the program from 2009 through the end of 2011. For information on New Jersey's withdrawal from RGGI, see http://rggi.org/design/history/njparticipation.

[6] CO_2 emissions account for approximately 99% of all GHG emissions from power plants. In 2011, electricity power plants accounted for about 33% of all U.S. GHG emissions. EPA, Draft 2012 U.S. Greenhouse Gas Inventory Report, February 2013.

[7] Electricity generating units that consume on-site more than 10% of the electricity they generate (on an annual basis) are not subject to the emissions cap. This provision applies to facilities (e.g., some refineries) that may generate electricity for their own use.

[8] For more details, see RGGI's website at http://www.rggi.org/design/overview/ regulated_sources.

[9] Based on 2009 data from World Resources Institute, Climate Analysis Indicators Tool (CAIT US) Version 5.0, 2012, at http://cait.wri.org.

[10] RGGI Memorandum of Understanding (MOU), Section 2 (G)(1), December 5, 2005. Subsequent amendments were made to the MOU, but not to this section.

[11] Safety-valve is a term of art in cap-and-trade parlance, but it can have a variety of meanings. Generally triggered by an established price in the emission allowance market, safety-valves may include (1) a set price alternative to making reductions or buying allowances at the market price (often described as a price safety-valve), (2) a slowdown in tightening the emissions cap, and (3) lengthening of the time allowed for compliance.

[12] For example, H.R. 2454 in the 111th Congress included a "strategic reserve" of allowances borrowed from future years. The reserve would be triggered at particular price points. California's cap-and-trade system has a similar mechanism. More information is available at http://www.arb.ca.gov/cc/capandtrade/capandtrade.htm.

[13] For related RGGI documentation, including the revised Model Rule, see http://www.rggi.org/design/ program_review.

[14] States from the Northeast and Mid-Atlantic regions began to discuss a cooperative effort to reduce carbon dioxide emissions in 2003. Subsequent meetings and workshops culminated in a Memorandum of Understanding (MOU) that was signed by most RGGI state governors in December 2005.

[15] See RGGI modeling results and analysis at http://www.rggi.org/design/program_ review/ materials-by-topic/ modeling; Environment Northeast, RGGI's Past and Future: Emissions Trends and Potential Reforms, 2012. Prior analyses include New York State Energy Research and Development Authority, Relative Effects of Various Factors on RGGI Electricity Sector CO_2 Emissions: 2009 Compared to 2005, Draft White Paper, November 2010; and Environment Northeast, RGGI Emissions Trends, June 2010.

16 The Energy Information Administration website provides a table listing the amount of CO_2 generated per unit of energy for different energy sources, at http://www.eia.doe.gov/ oiaf/1605/coefficients.html.

[17] In its original Memorandum of Understanding (December 20, 2005), RGGI states agreed to conduct a "comprehensive review" of the RGGI program in 2012.

[18] Between 2009 and 2013, the emissions cap exceeded or will exceed actual emissions, providing an opportunity for entities to obtain more allowances than they need to meet current compliance obligations. These allowances can be purchased and held (i.e. banked) for future use.

[19] New York State Energy Research and Development Authority (prepared for RGGI Inc.), Relative Effects of Various Factors on RGGI Electricity Sector CO_2 Emissions: 2009 Compared to 2005, Draft White Paper, November 2010.

[20] See RGGI Inc., Investment of Proceeds from RGGI CO_2 Allowances, February 2011 and Environment Northeast, Economy-wide Benefits of RGGI: Economic Growth through Energy Efficiency, March 2011.

[21] Environment Northeast, Economy-wide Benefits of RGGI: Economic Growth through Energy Efficiency, March 2011.

[22] Possible measures included carbon adders, procurement emissions rates, or load-based caps. See RGGI Emissions Leakage Multi-State Staff Working Group to the RGGI Agency Heads, Potential Emissions Leakage and the Regional Greenhouse Gas Initiative (RGGI), (March 2008), at http://www.rggi.org/about/documents.

[23] RGGI, RGGI 2012 Program Review: Summary of Recommendations to Accompany Model Rule Amendments, February 2013, at http://www.rggi.org/design/program_ review/ materials-by-topic/program-elements.

[24] At least one of the RGGI states is scheduled to increase its percentage of allowances offered through auction. For details, see Environment Northeast, RGGI Auction Tracker: State Allocations and Spending Plans, September 2012.

[25] For more information, see RGGI Inc., Investment of Proceeds from RGGU CO_2 Allowances, February 2011; and Environment Northeast, RGGI Auction Tracker: State Allocations and Spending Plans, March 2011.

[26] CRS Report R40556, Market-Based Greenhouse Gas Control: Selected Proposals in the 111th Congress, by Larry Parker, Brent D. Yacobucci, and Jonathan L. Ramseur.

[27] For information on other auction formats, see Charles Holt et al., Auction Design for Selling CO_2 Emission Allowances Under the Regional Greenhouse Gas Initiative, 2007, prepared for RGGI Working Group staff.

[28] RGGI states proposed that the 2014 reserve price will be \$2.00, increasing by 2.5% each year thereafter.

[29] CRS Report R42731, Carbon Tax: Deficit Reduction and Other Considerations, by Jonathan L. Ramseur, Jane A. Leggett, and Molly F. Sherlock.

[30] In a cost-effective emissions trading program, the allowance price should mirror (or closely follow) the marginal cost of emission reduction—that is, the cost of reducing the last, most expensive ton. An effective auction should help identify the allowance price that is near to the marginal cost of reduction. See, for example, Charles Holt et al., Auction Design for Selling CO_2 Emission Allowances Under the Regional Greenhouse Gas Initiative, 2007, prepared for RGGI Working Group staff.

[31] See RGGI allowance allocation data, at http://www.rggi.org/market/ CO_2_ auctions/allowance_ allocation.

[32] For more discussion of these issues, see CRS Report RL34502, Emission Allowance Allocation in a Cap-and-Trade Program: Options and Considerations, by Jonathan L. Ramseur.

[33] RGGI documents are available at http://www.rggi.org/.

[34] See RGGI Model Rule, issued August 15, 2006, p. 42; and RGGI Memorandum of Understanding, Section G(1), signed by participating state governors December 20, 2005.

[35] For details, see Environment Northeast, RGGI Auction Tracker: State Allocations and Spending Plans, September 2012.

[36] If allowed as a compliance option in a cap-and-trade system, offsets have the potential to provide considerable cost savings and other benefits. However, offsets have generated considerable controversy, primarily over the concern that illegitimate offsets could undermine the ultimate objective of a cap-and-trade program: emission reduction. For more discussion see CRS Report RL34436, The Role of Offsets in a Greenhouse Gas Emissions Cap-and-Trade Program: Potential Benefits and Concerns, by Jonathan L. Ramseur.

[37] For example, H.R. 2454 (Waxman-Markey) in the 111th Congress, would have allowed offsets to satisfy 27% of a facility's compliance obligation in 2016.

[38] RGGI states proposed to expand this offset project type to include "reforestation, improved forest management, or avoided conversion."

[39] See CRS Report RL34436, The Role of Offsets in a Greenhouse Gas Emissions Cap-and-Trade Program: Potential Benefits and Concerns, by Jonathan L. Ramseur.

[40] See http://www.rggi.org.

[41] See CRS Report R41212, EPA Regulation of Greenhouse Gases: Congressional Responses and Options, by James E. McCarthy.

[42] See e.g., Franz Litz et al., What's Ahead for Power Plants and Industry? Using the Clean Air Act to Reduce Greenhouse Gas Emissions, Building on Existing Regional Programs, World Resources Institute, 2011.

[43] Calculated by CRS using 2011 data from the Bureau of Economic Analysis, at http://www.bea.gov.

[44] For example, 16% of RGGI's in-state electricity comes from coal and 35% comes from nuclear energy. In the United States as a whole, approximately 44% of electricity is generated from coal and 20% from nuclear energy. See Energy Information Administration data.

In: The Regional Greenhouse Gas Initiative ISBN: 978-1-62808-483-2
Editor: Serge Chauvin

Chapter 2

EMISSION ALLOWANCE ALLOCATION IN A CAP-AND-TRADE PROGRAM: OPTIONS AND CONSIDERATIONS*

Jonathan L. Ramseur

SUMMARY

When designing a cap-and-trade program, one of the more controversial and challenging questions for policymakers is *how, to whom, and for what purpose* to distribute the emission allowances. Regarding the *method* of distribution to covered sources, policymakers could (1) sell the allowances through an auction process, (2) allocate the allowances at no cost to covered sources, (3) provide allowances to non-covered sources who would, in turn, sell them to covered sources, or (4) use some combination of these methods. Although the emission allocation method *would not* affect the environmental integrity of the cap-and-trade program, the selected allocation strategy could have considerable consequences.

Using auctions as a distribution method could avoid certain concerns that are likely to occur if covered sources receive all (or most) of the allowances at no cost: (1) consumers in different electricity markets may face inequitable price increases; (2) a weak price signal for electricity may be sent in areas with the most carbon-intensive fuel portfolios; and

* This is an edited, reformatted and augmented version of CRS Report for Congress, Publication No. RL34502, dated June 2, 2008.

(3) no-cost allowances may overcompensate covered sources. In addition, auction revenues offer a unique opportunity to reduce the overall costs of the emissions program. Several economic studies indicate that if used in the most efficient manner, overall costs could be minimized by almost 50%.

A greenhouse gas (GHG) emission cap-and-trade program would create a valuable new commodity: the GHG emission allowance. EPA estimates that allowance value could potentially account — in aggregate — for tens or hundreds of billions of dollars each year. When distributing this value, policymakers would face a choice between minimizing the costs imposed on the entire economy, minimizing the expected burden on specific parties, or supporting a range of climate- or nonclimate-related policy objectives.

For example, Congress may consider providing transition assistance to carbon-intensive industries. Studies have estimated profits could be maintained in the energy production and electricity generation sectors, if approximately 20% of allowances were provided to those sectors at no cost. Members may also consider allotting allowance value to consumers, particularly low-income households, who are expected to bear the majority of the compliance costs via higher energy prices. Another option would involve distributing the allowance value to support various objectives: technology development, energy efficiency improvements, biological sequestration, climate change adaptation efforts, or non climate-related purposes, such as deficit reduction. Of these objectives, technology advancement is arguably the most crucial in terms of mitigation. Moreover, deployment of new technologies could potentially lower the overall costs of the program.

Although many of the proposals in the 110th Congress (e.g., S. 2191, S. 1766, and S. 3036) would employ an auction to some degree, none of the bills specifies the design of the auction. Congress may want to consider including specific design elements in legislative text, particularly auction frequency and whether or not the auction should have a reserve price, and if so, at what level.

INTRODUCTION

Climate change issues have generated interest and debate over the past two decades. In 1992, the United States ratified the United Nations Framework Convention on Climate Change (UNFCCC). Arguably, in recent years the primary issues under debate have shifted from science — such as the role of greenhouse gas (GHG) emissions from human activities[1] — to policy. For instance, a 2005 Sense of the Senate Resolution on climate change[2] stated:

> It is the sense of the Senate that Congress should enact a comprehensive and effective national program of mandatory, market-based limits and incentives on emissions of greenhouse gases that slow, stop, and reverse the growth of such emissions at a rate and in a manner that, No. 1, will not significantly harm the U.S. economy and, No. 2, will encourage other action and key contributors to global emissions.

In Congress, Members introduced several bills that would establish a market-based,[3] mandatory GHG emission reduction program.[4] Most of these proposals would establish some type of cap-and-trade system to regulate GHG emissions.[5] For a brief primer on cap-and-trade systems, see Appendix A.

In designing a cap-and-trade program, one of the more controversial and challenging questions for policymakers is how, to whom, and for what purpose to distribute the emission allowances. Concerning the question of how to distribute allowances, policymakers could (1) sell the allowances through an auction process (2) allocate the allowances at no cost to covered sources, (3) provide allowances to non-covered sources, who would, in turn, sell them to covered sources via the emissions trading market, or (4) use some combination of these methods.

Regardless of the method of distribution, emission allowances would have monetary value in a carbon-constrained regime, such as a cap-and-trade program. If an auction is used, policymakers could distribute auction revenues to a wide range of parties to support various policy objectives. Likewise, policymakers could allot allowances at no cost to non-covered entities — e.g., federal or state agencies, among others — to promote the same (or different) objectives.

By addressing the question of how, to whom, and for what purpose to distribute the emission allowances, policymakers would craft an allocation strategy. The strategy *would not* affect the environmental integrity of the emissions cap.[6] In addition, covered entities would generally face the same emission reduction decisions under either allocation strategy.[7] A "common misconception" is that if covered sources receive allowances at no cost, the sources would behave differently from sources who purchased allowances through an auction.[8] Economists point out "free allowances"[9] have value, and when covered entities submit an allowance for compliance purposes, the entities forgo the opportunity[10] to sell the unused allowance in the emissions trading market.[11]

The first two sections of this report discuss the primary emission allowance distribution *methods*: auctions and no-cost distribution to covered

sources. These sections examine the potential benefits and concerns of these allocation mechanisms. The final section identifies different options and policy considerations for Congress when determining to whom and for what purpose to distribute the value of the emission allowances. The allocation strategy would have substantial consequences for the cost of the cap-and-trade program and how the costs are apportioned.

AUCTIONS

In recent years, the use of auctions to allocate emission allowances has generated considerable interest.[12] Several of the cap-and-trade proposals from the 110th Congress — including S. 2191, which was reported from the Senate Environment and Public Works Committee on May 20, 2008 — would use auctions to allocate an increasing percentage of the cap's emission allowances.[13]

This section describes the potential benefits that auctions may provide, if used to distribute allowances to covered sources in a cap-and-trade program. In addition, this section discusses auction design issues and considerations for policymakers.

Implementation Benefits

In general, the *concept* of an auction is relatively simple to understand.[14] Auctions would allow the market to determine which entities receive emission allowances and at what price: parties placing the highest value on the allowances would receive them. With this allocation method, policymakers would be relieved of the responsibility to make distribution decisions, a process that might be described as picking winners and losers.[15] For this reason, auctions are generally considered to be the most transparent mechanism for distributing allowances.

In addition, in a free allocation format parties would have strong incentives to seek increasing shares of the overall allowance allotment.[16] Parties with resources available for such efforts may have an advantage. An auction system would eliminate this behavior.

Polluter Pays Principle

Requiring emission sources to purchase emission allowances would support the "polluter pays" principle. In a general environmental policy context, the polluter pays principle holds that pollution costs should be borne by the polluting facility or industry, not society at large. To accomplish this objective, pollution costs should be included in the overall price of a good. Proponents of the polluter pays notion would likely argue that if products are priced to reflect environmental costs — air pollution, land use, GHG emissions — demand for these products may decline.

Advocates of polluter pays maintain that the environment and the services it provides are a shared public good. Under this framework, facilities should have to pay for the right to pollute (i.e., emit GHGs). If allowances are provided at no cost to emission sources, the polluter pays principle would be violated.

Potential Minimization of Costs on Society

If Congress decides to use an auction to distribute emission allowances to covered sources — as opposed to providing allowances to covered sources at no cost — the auction revenues could be used to substantially minimize the overall costs on society of the cap-and-trade program.

Economic studies have found that, if revenues are used *in the most economically efficient manner*, the overall costs imposed by a cap-and-trade program could be reduced by approximately 50%.[17] Economists maintain that the most economically efficient application of revenues would be to offset reductions in taxes on desirable activities, such as employment or personal income.[18] The opportunity to use allowance value in this manner and thus minimize overall costs to this extent is unique to the auction mechanism.[19] However, many observers argue that applying auction revenues in this fashion is politically unlikely.[20] Other potential uses of the auction revenues may or may not generate *overall* economic cost savings. These options are discussed later in this report.

Avoidance of Economic Concerns from No-Cost Distribution

Auctions would avoid several of the undesired economic effects that are likely to occur if allowances are provided to covered sources at no cost. These concerns are discussed in greater detail later in the no-cost distribution section. In brief, they include:

- Inefficient and inequitable price signals in the electricity sector;
- Potential overcompensation to covered sources; and
- Challenges with allotting allowances to new and retiring sources.

Auction Design Issues

Although many of the cap-and-trade proposals in the Congress would employ an auction to some degree, none of the bills specifies the design of the auction. A recent study that examined auction design issues for the Regional Greenhouse Gas Initiative (RGGI) found that "careful attention to auction design can be critical to an auction's success in achieving the goals specified for the auction."[21]

Design Considerations

The success of an auction is typically measured by both its efficiency and revenue generation.[22] In an emission auction context, efficiency is achieved when the parties that receive the allowances are the parties that place the most value on the allowances. Other attributes of an auction that may be used to measure its success include:

Price Discovery

In a cost-effective emissions trading program, the allowance price should mirror (or closely follow) the marginal cost of emission reduction — i.e., the cost of reducing the last, most expensive ton. An effective auction should help identify the allowance price that is near to the marginal cost of reduction.[23]

Protection against Market Manipulation

Auctions should discourage or prohibit bidding behavior that would create inefficient outcomes in the market. For example, collusion among bidders may artificially lower the allowance price. Another concern is hoarding, in which one party makes speculative bids above the competitive price, in order to capture a disproportionately large percentage of the allowances.

Minimize Transaction Costs

Substantial administrative or transaction costs could reduce the cost-effectiveness of using an auction. Moreover, high transaction costs could place smaller firms at a disadvantage.

Transparency and Fairness

The rules should be readily available to all parties and should not favor certain participants.

Design Options

Certain auction designs may provide advantages or disadvantages, depending on the auction's primary objective. For example, some auction designs in certain contexts may favor revenue generation; others may be more efficient in terms of matching the market price.

Policymakers may undertake further study before specifying the particular auction logistics. Although economic studies have examined the performances of different auction formats in other contexts, "relatively few papers have examined the relative merits of each of these auction forms in multi-unit [e.g., emission allowances] auctions."[24] One option for Congress would be to direct an implementing agency to devise the most appropriate auction format, based on the ranking of objectives provided by Congress.

Policymakers may consider various auction designs . In general, auction designs are distinguished by (1) the number of rounds for bidding — generally one round (often called "sealed bid") versus multiple rounds; and (2) whether there is a uniform price or individual price ("discriminatory" price) for each buyer. Examples of auction designs with different combinations of these two characteristics include the following:

Discriminatory Price, Sealed-Bid Auction

This type of auction is used in EPA's SO_2 emission trading program.[25] In this system, parties submit a sealed bid, containing multiple offers to purchase a set number of allowances at certain prices. The implementing agency opens the bids and distributes allowances, starting with the highest offer, until the supply is exhausted. For example, consider a hypothetical auction, in which the supply of allowances is 20 units and the highest bidder offered $100 per allowance for 15 allowances, and the second highest bidder offered $90 per allowance for 10 allowances: the highest bidder would receive 15 allowances for $100/allowance; the second highest bidder would receive 5 allowances at $90/allowance.

Uniform-Price, Sealed-Bid Auction

This type of auction is similar to the above format — discriminatory price, sealed-bid — with one major difference: the price paid by all bidders is the highest rejected bid (i.e., the second-highest bid). Using the above scenario, the highest bidder would receive 15 allowances at $90/allowance, and the second-highest bidder would receive 5 allowances at $90/allowance. Ireland used this design to implement its auction for the EU ETS.[26] In addition, in a study prepared for RGGI officials, researchers recommended using this approach.[27]

Uniform-Price, Multi-Round (English Clock) Auction

In an emission allowance auction using this format, the auctioneer would post a allowance price and parties would offer the quantity they are willing to buy at the posted price. The first posted price would be set at a low level, so that demand would exceed supply. The auctioneer would continue posting higher prices at set time intervals, until demand is less than (or equal to) the allowance supply. The posted price that produces this outcome would become the allowance price for all bidders. Virginia used this auction type to sell nitrogen oxide emission allowances pursuant to the "NO_x SIP Call."[28]

Reserve Price

One issue that arguably transcends auction design considerations is whether or not the auction should have a reserve price, and if so, at what level. In an auction, a reserve price is a price set by the seller, below which the seller refuses to part with the item for sale. In a large volume, multi-unit auction that is expected to have substantial participation (i.e., high demand for the items for sale), a reserve price would all but guarantee a revenue stream. In a cap-and-trade allowance emissions auction, a reserve price would operate much like a minimum tax or price floor.

A reserve price may address certain logistical concerns, such as bidder collusion, that are often associated with auctions. In addition, a reserve price may provide assurance to parties making emission reductions that the reductions will have some value in the allowance market. For example, if a covered source can expect a reserve price to be set at a certain level (e.g., $10/ton), and the source makes multiple reductions, each at a per-ton cost below the expected reserve price, the source can have confidence that its efforts will be cost-effective.

The authors of the RGGI auction study recommended that RGGI participants set a reserve price when conducting allowance auctions, concluding:

> A compelling justification for a reserve price can be found in the academic literature and from previous experience with auctions, and the reserve price would help the auction achieve the criteria [e.g., the design criteria discussed above] set out in this report.[29]

Because a reserve price (if established) could influence revenue flows from an auction, Congress may consider addressing this issue specifically in legislative text, rather than leave this matter open for interpretation to an implementing agency.

Auction Frequency

EPA's SO_2 emissions trading program holds annual auctions to distribute a small percentage of allowances. Likewise, S. 2191 would direct the implementing agency to conduct annual auctions. However, policymakers may consider holding more frequent auctions (e.g., every quarter).

More frequent auctions could provide several benefits, both for covered sources and to the efficiency of the program. More auctions would give covered sources more flexibility to incorporate unanticipated events — e.g., higher electricity demand due to warmer than expected temperatures. If auctions were held more frequently, the allowances sold would be in smaller lots. This may help facilities, particularly smaller operations, maintain cash flow.

In terms of efficiency, smaller, more frequent auctions would likely reduce the potential for parties to manipulate the market (e.g., from speculative hoarding). More auctions may increase market liquidity by making allowances available for purchase in more frequent intervals.

The potential downside to having multiple auctions per year is that covered sources may face additional transaction costs. However, the authors of the RGGI auction study stated:

> past experience suggests that a significant proportion of the administrative cost of holding auctions is related to the initial set-up ... and that incremental costs of repeating a particular auction type will be low in comparison to these initial costs.[30]

After considering the "costs, risks, and benefits," the study authors recommended that participating RGGI states use a quarterly auction to allocate emission allowances.[31]

No-Cost Distribution to Covered Sources

Emission trading programs, in the United States[32] and abroad,[33] have generally distributed the vast majority of allowances at no cost to sources directly subject to a cap. In recent years, however, support for auctions has gained momentum. This momentum likely reflects a better understanding of the benefits of using auctions, as well as increased scrutiny of the effects of distributing allowances at no cost.[34] This section discusses the potential benefits and concerns of allotting allowances to covered sources at no cost.

Potential Benefits

Mitigation of Disproportionate Costs

The primary argument in support of no-cost distribution is that carbon-intensive industries are expected to face disproportional costs under a carbon-constrained system. These industries maintain they should receive compensation (i.e., free allowances) for the financial losses imposed by the cap-and-trade program. The financial losses may lead to loss of jobs in particular industries. The compensation may be considered a form of transition assistance for industries and industry employees most impacted by a GHG emissions cap.

This argument is perhaps stronger for industries that may have a more difficult time including the costs of emission reduction in the price of their products. For example, certain U.S. industries may be more vulnerable to foreign competition, especially if their competitors are located in nations without GHG emissions caps. For these industries, increasing the price of their materials (to reflect the cost of emissions abatement) may entail a comparative disadvantage. Moreover, if foreign competitors in these industries increase their market share as a result of a U.S. capand-trade program, the foreign facilities (in uncapped economies) are likely to increase their GHG emissions. This potential scenario is described as emissions leakage, a constant concern in climate change policy.

In other economic sectors, particularly the electricity generation sector, that do not face foreign competition, facilities are expected to pass along the vast majority of the emission reduction costs. This would entail higher prices for consumers, which includes businesses and households. However, price increases would likely reduce consumer demand,[35] potentially lowering the profits of carbon-intensive industries.[36]

Political Feasibility

Cap-and-trade programs, both domestic and international, have usually provided allowances to covered sources at no cost; this free allocation to covered sources is arguably a means to garner support for an emissions reduction program. Moreover, industries may prefer to receive allowances at no cost rather than compete for a share of auction revenues: a transfer of free allowances may be more "politically secure than government promises of compensation from auction revenues."[37]

Concerns

Undesirable Effects in the Electricity Sector

As the electricity sector accounts for the largest percentage (34% in 2006)[38] of GHG emissions in the United States, it would play a major role in the effectiveness of a cap-and-trade program. There is concern that if allowances are distributed to electric utilities at no cost, the electricity consumers who purchase electricity under a price-regulated structure would receive a price signal that is weaker than the signal received by consumers in unregulated (or competitive) markets.[39] In contrast, an auction distribution system would enable electricity generators, in both price-regulated and competitive markets, to send a comparable price signal to consumers.[40]

In the United States, the price consumers pay for electricity may be determined by a state regulatory body — often described as cost-of-service regulation — or the price may be subject to market forces — often described as deregulated or competitive. In general, the regulatory structure varies by the type of facility and/or the state in which the electricity is generated. In 2007, the more traditional, price-regulated electric utilities generated approximately 60% of the total net electricity generated in the United States.[41] The remaining 40% of electricity was generated by facilities that are unregulated in terms of their ability to set a price for the electricity they generate.

A comparison between Figure 1 and Figure 2 demonstrates the difference in price signals that the two regulatory frameworks would send.[42] These figures present a distribution of electricity price changes that consumers could expect under a cap-and-trade system that included the electricity sector.[43] For example, under an auction system (Figure 1), a large percentage of consumers would face an increase of $8/Mwh; approximately equal numbers of consumers would face either higher or lower prices.

In Figure 1, the change in electricity price for both "regulated" and "competitive" regions is fairly symmetrical. The (rise-and-fall) shape of the column heights, which is generally identical for both the price-regulated and competitive regions, reflects the different electricity-generating fuel portfolios that exist throughout the country. Regions with higher carbon content fuel portfolios[44] are expected to experience higher electricity prices under a cap-and-trade system.

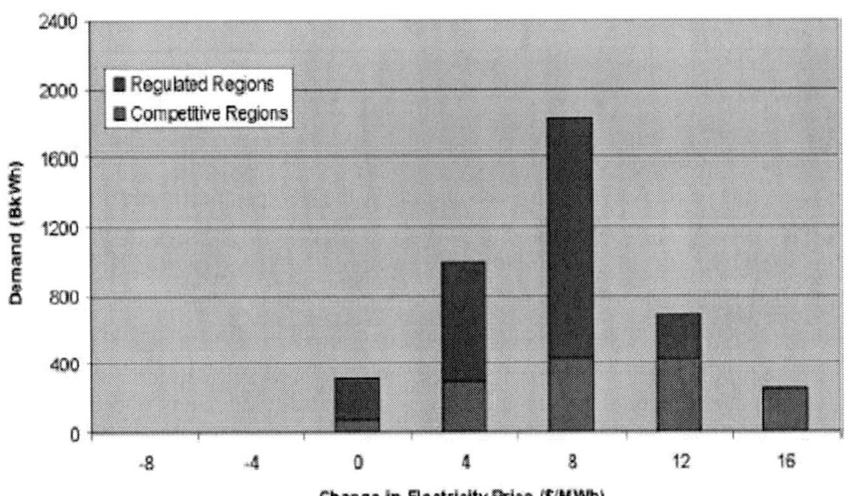

Source: Dallas Burtraw (Resources for the Future), Prepared Testimony before the House Select Committee on Energy Independence and Global Warming, January 23, 2008.

Figure 1. Change in Electricity Price by Region: Allowances Distributed to Covered Sources with Auction.

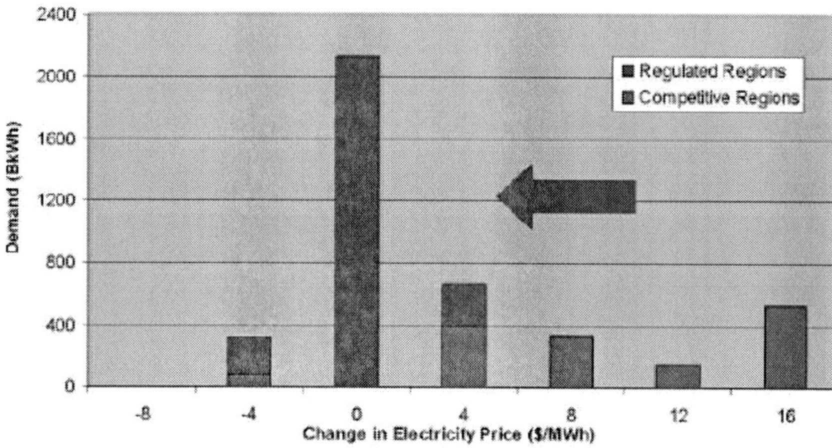

Source: Dallas Burtraw (Resources for the Future), Prepared Testimony before the House Select Committee on Energy Independence and Global Warming, January 23, 2008.

Figure 2. Change in Electricity Price by Region: Allowances Distributed to Covered Sources at No Cost.

However, when covered sources receive allowances at no cost (Figure 2), consumers in price-regulated and competitive regions experience dramatically different price changes. Note the asymmetrical shape of the columns, as compared to those in the previous figure. In most of the price-regulated regions, the electricity price would remain the same (or decrease), while the price would increase in most of the competitive regions.

The different consumer impacts identified in Figure 1 and Figure 2 result from the dissimilar market structures — price-regulated versus competitive — that determine the price of electricity for U.S. consumers. The different impacts reflect the electric- generating facilities' varied abilities to pass through all types of costs to consumers in the form of higher electricity prices. In an auction, the costs to utilities would include both the costs of mitigation and the costs of purchasing allowances. Under a no-cost distribution system, utilities would have mitigation costs and opportunity costs associated with the allowances (discussed above).

If policymakers auction allowances to electric utilities, both price-regulated and competitive-market utilities would include the cost of purchasing emission allowances in the price of electricity. However, if allowances are distributed at no cost to utilities, only competitive-market utilities would be able to pass along their opportunity costs.

Because price-regulated facilities would not be able to pass through the opportunity costs associated with free allowances, consumers in these areas would effectively receive the benefit of the "free" allowances in the form of stable or lower electricity bills (Figure 2). Consumers in competitive areas would not receive the benefit of "free allowances" and would thus face disproportionate price increases. This consequence would erode the effectiveness of the cap-and-trade program, because consumers would likely not receive a price signal that is strong enough to encourage conservation or energy efficiency improvements.

The effects of this inefficient outcome may be magnified, because approximately 75% of coal-fired electricity was generated by price-regulated utilities in 2007.[45] Thus, consumers that utilize more-carbon intensive electricity would face a weaker price signal than consumers using less carbon-intensive electricity.[46]

The price disparities that consumers with comparable fuel portfolios would experience under different electricity regulatory structures would be both unfair and inefficient. An auction would eliminate both of these concerns. Consumers in price-regulated and competitive regions (with similar carbon-intensive electricity profiles) would experience more equitable impacts. Moreover, the costs of the cap-and-trade program would be included in the price of electricity in each market structure. This result is necessary if the carbon price is to modify consumer behavior: e.g., spur energy conservation efforts or the installation of more energy efficient technologies.

Overcompensation to Covered Entities

If covered entities receive, at no-cost, GHG emission allowances in proportion to their emissions, there is concern that the recipients would be overcompensated for the compliance costs imposed by a cap-and-trade program.[47] Depending on the percentage of emission allowances auctioned, an auction could avoid overcompensation.

This potential outcome is a function of several factors. First, the aggregate value of the GHG emission allowances is expected to be substantially greater than the aggregate costs of making emission reductions pursuant to the emissions cap.[48] (See Figure 3 and surrounding discussion below.) Second, if covered sources receive allowances at no cost, they would retain the benefits of these allowances, which are essentially a form of currency.[49] Although it may be counterintuitive, covered entities are expected to raise the price of their products, even if entities receive allowances at no cost.[50] Thus, covered sources would receive the financial benefit of the allowances and the gains associated with higher prices.[51] These benefits are often described as "windfall profits."[52]

A windfall profit result has been observed in cap-and-trade models[53] and in the largest existing cap-and-trade program: the European Union's (EU) Emissions Trading Scheme (ETS). The EU ETS established a cap-and-trade program for power plants and certain carbon-intensive industries, allocating virtually 100% of the allowances at no cost to covered entities between 2005 and 2007 ("Phase 1"). One study estimated that power plants in the United Kingdom received windfall profits of 800 million euro (approximately $1 billion)[54] per year;[55] a separate study estimated windfall profits for power plants in the Netherlands at 300 million to 600 million euro ($378 million - $750 million)[56] per year.[57] However, the EU ETS Phase 1 allowance price (for reasons beyond the scope of this report) dropped dramatically in April 2006 and never recovered.[58] Thus, the windfall profits were only generated in the first year (2005) of the program.

Nevertheless, a 2008 study estimated that European power plants are expected to continue to receive windfall profits during the second phase (2008-2012) of the EU ETS. For example, German plants are projected to generate the most windfall profits: between 14 billion and 22 billion euro ($21 billion - $33 billion)[59] over that time period.[60] In recognition of this projection, the European Commission (EC) has proposed to address this issue, stating: "taking into account their ability to pass through opportunity costs, full auctioning should be the rule from 2013 onwards for the power sector." In addition, the EC has proposed that free allocation in other sectors would gradually phase out, so that by 2020, auctions would distribute 100% of the allowances.[61]

Treatment of New and Retiring Sources

An auction distribution format would address economic inefficiencies and concerns of fairness regarding new emission sources and facilities that are near retirement. A recurrent auction creates a level playing field for these two categories of sources. In contrast, a distribution strategy that allots allowances at no cost based on prior year emissions ("grandfathering") could provide a considerable advantage to existing facilities. A free allowance effectively subsidizes currently operating facilities, which may be using outdated, inefficient technologies.[62] Moreover, if entities receive allowances at no cost, the financial gain imparted in the allowance could serve as an incentive to extend the facility's operation beyond a time that would otherwise be efficient to cease operations.[63]

Distribution of Allowance Value: Options and Considerations

By limiting the annual number of emission allowances available for compliance purposes, a cap-and-trade system creates emission allowances. Emission allowances would become a valuable new commodity, potentially accounting — in aggregate — for tens or hundreds of billions of dollars (Table 1 and surrounding discussion). To whom and for what purpose the value of allowances is distributed would affect (1) the overall cost to society of a cap-and-trade program and (2) which parties bear the costs of the program. Policymakers would face a choice between minimizing the costs imposed on the entire economy (society's costs) or using the allowance value for other purposes.[64] The latter choice covers a range of options. Congress could choose to provide assistance to specific industries or groups, or choose to distribute the allowance value to support various objectives. To provide an example, Appendix B of this report identifies the allowance and auction revenue distribution strategies proposed by S. 2191.

The first part of this section provides an overview of some key concepts and an estimate of the total allowance value that may be available for distribution in a capand-trade program. The second part examines the range of options for distributing allowance value. This is followed by a discussion of policy considerations.

Overview and Estimate of Allowance Value

Compliance Costs Versus the Value of Emission Allowances

A cap-and-trade program would impose costs: covered entities would comply by reducing their own emissions, purchasing emission reductions (credits) from other covered entities, or (if allowed) buying offsets from non-covered sources that have reduced, avoided, or sequestered emissions.[65] The combined costs of these activities are the "compliance costs" of the cap-and-trade program.[66]

The compliance costs are different from the aggregate value of emission allowances. In the early years of a cap-and-trade program, the aggregate value of allowances would likely dwarf the costs of making (or finding) emission reductions.

Figure 3 depicts this contrast: the area under the emissions cap curve represents the number of allowances; the area between the emissions cap curve and the business-asusual curve represents the required reductions. The former area will be larger as long as the emission reduction target is less than 50% of the emissions baseline.[67]

The compliance costs represent the sum of the costs of each ton of reduction, and the cost of each reduced ton will vary. For example, some projects may present "low-hanging fruit" reduction opportunities, whereas investments in extra capital (e.g., carbon capture technology) may represent a more expensive reduction option.

Consider a simplified example:[68] policymakers set a 7-ton cap on an economy that currently emits 10 tons. The 3 tons that must be reduced cost $1, $5, and $10, respectively. The cost of the last, most expensive ton — the marginal cost — is $10. Because the marginal cost largely establishes the market price of emission allowances in a cap-and-trade, each emission allowance has a value that is approximate to the marginal cost.[69] Therefore, in this example, the value of the allowances would equal $70, but the compliance costs would be only $16.

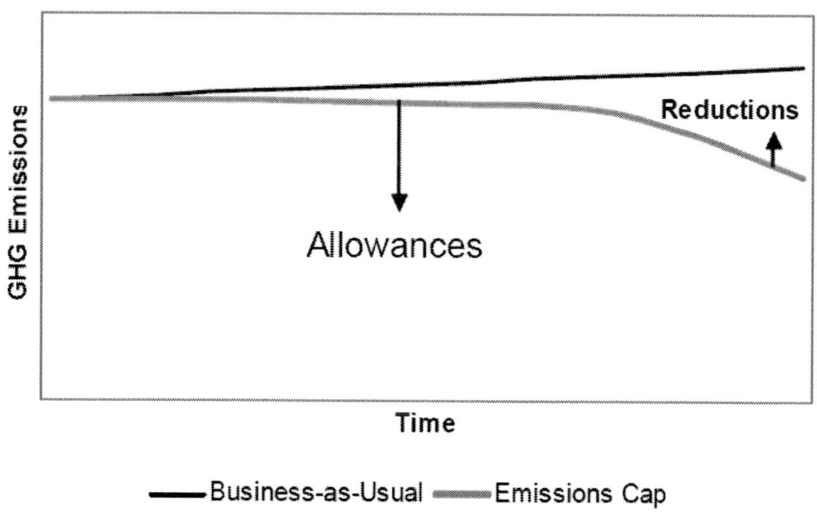

Source: Prepared by CRS. The concept for the figure comes from National Commission on Energy Policy, *Allocating Allowances in a Greenhouse Gas Trading System* (2007).

Figure 3. Number of Allowances vs. Number of Reductions.

Estimates of Allowance Value

Whether or not Congress sells the allowances through an auction (and generates revenues) or distributes the allowances at no cost, the allowances would have monetary value. To put the allowance value in context, consider the Lieberman-Warner Climate Security Act of 2008 (S. 2191), an "economy-wide"[70] cap-and-trade proposal that was reported out of the Senate Environment and Public Works Committee May 20, 2008. At the request of Senators Lieberman and Warner, EPA prepared an economic analysis of provisions of S. 2191.[71] One of the primary results of the analysis is the estimated price range for emission allowances. From these price estimates, potential auction revenues can be identified, because the legislation specifies the percentage of allowances to be auctioned in each compliance year. Table 1 lists the estimated annual action revenues that would be generated under provisions of S. 2191. In addition, the table provides an estimate of revenues that would be generated if 100% of the emission allowances were auctioned. This latter estimate represents an approximation of the aggregate value of the emission allowances in a given year.

Table 1. Estimates of Auction Revenue under the Framework of the Lieberman-Warner Climate Security Act of 2008 (S. 2191)

	S. 2191				**S. 2191 w/ 100% Auction**
Year	**Total Allowances Available ($mtCO_2$-e)**	**Percentage of Allowances Auctioned**	**Estimated Emission Allowance Price Range (\$/$mtCO_2$-e)**	**Estimated Annual Auction Revenue (\$ billions)**	**Estimated Annual Auction Revenue (\$ billions)**
2015	5,456	29.5%	\$29 - \$40	**\$47 - \$64**	**\$158 - \$218**
2020	4,924	36.5%	\$37 - \$51	**\$67 - \$92**	**\$182 - \$251**
2025	4,392	48.5%	\$48 - \$65	**\$101 - \$139**	**\$211 - \$285**
2030	3,860	62.8%	\$61 - \$83	**\$147 - \$201**	**\$235 - \$320**
2035	3,328	69.5%	\$77 - \$106	**\$179 - \$245**	**\$256 - \$353**
2040	2,796	69.5%	\$98 - \$135	**\$191 - \$263**	**\$274 - \$377**
2045	2,264	69.5%	\$125 - \$173	**\$197 - \$272**	**\$283 - \$392**
2050	1,732	69.5%	\$159 - \$220	**\$192 - \$265**	**\$275 - \$381**

Source: Prepared by the Congressional Research Service with estimates of annual emission allowance prices provided by U.S. EPA, *EPA Analysis of the Lieberman-Warner Climate Security Act of 2008* (2008), p. 27.

Note: The price range represents the results of two separate models. For more information, see Appendix 1 of EPA's analysis of S. 2191. In addition, the figures in the "Total Allowances Available" column are from the reported version of S. 2191 (May 20, 2008) and do not include proposed changes for deficit reduction purposes.

To put the auction revenues/allowance value in context, consider the federal net tax revenue from the three largest revenue sources for Fiscal Year 2007:[72]

- individual income tax: $1,118 billion;
- employment taxes: $838 billion;
- corporate income taxes: $368 billion.

Options for Allowance Value Distribution

Provide Transition Assistance to Carbon-Intensive Industries

Several economic studies have estimated the percentage of allowances (a comparable amount of auction revenues could also be used) that would provide compensation for projected profit losses to specific carbon-intensive industries. The findings include the following:

- Goulder found that 13% of the emission allowances would compensate the *fossil fuel producing industry* (coal mining, oil and natural gas extraction) for lost profits.[73]
- Burtraw and Palmer concluded that the *electric-generating industry's* estimated profit losses could be offset with 6% of the emission allowances.[74]
- Smith and Ross estimated that 21% of the allowances would compensate the combined losses of primary energy producers and electric utilities (i.e., a combination of the sectors examined in the other two studies).[75]

Although these studies analyzed the net effects to certain economic sectors, there are likely to be winners and losers within the sectors, particularly in electricity generation. For instance, some facilities, such as coal-fired plants, are expected to see greater losses, while others — hydroelectric or renewable energy plants — may see a gain in profits.[76]

The above estimates consider that allowances would be provided to covered sources in perpetuity. However, some cap-and-trade proposals in the 110th Congress (e.g., S. 2191 and S. 1766) would generally phase-out free allocation as the percentage of allowances to be auctioned increases. As such, the initial percentage of allowances provided at no cost in these bills is higher than estimates that provide for indefinite distribution at a set percentage. For example, during the first five years of the program established by S. 2191, fossil fuel-fired power plants would receive 19% of the emission allowances at no cost.[77] This percentage would gradually decline to zero by 2031.[78]

Offset Reductions in Distortionary Taxes

Economic theory generally supports a tax policy that would reduce taxes on favored activities (increased employment or personal income) and increase taxes on less desirable behavior (increased pollution).[79] Auction revenues could be used to offset reductions in the taxes that apply to desirable activities.

Using auction proceeds in this manner has been described as yielding a double-dividend: (1) reduced GHG emissions and (2) reduced market distortions from the taxes on desirable behavior. In the early 1990s, some economists suggested that the double-dividend effect would be strong enough to keep the overall costs to society relatively small or even negative.[80] However, more recent economic studies indicate that the costs imposed by the cap-and-trade program could act as an additional tax,[81] which would most likely (at least in the United States)[82] exceed the benefits of revenue recycling.

Several economic studies have estimated the cost savings to society that auction revenues could provide.[83] A cap-and-trade program is expected to impose an economic cost on society.[84] A frequently cited study found that auctioning allowances could reduce the projected costs between 21% and 47%.[85] The range of potential cost savings reflects different uses of the auction revenues. If policymakers were to distribute the revenues to U.S. households in "lump-sum" payments — e.g., increase the standard tax deduction or mail payments to households — the cost savings would be on the lower end of the spectrum. Alternatively, if Congress decided to use the revenues to reduce taxes on labor or investment, society's costs would be minimized. Figure 4 shows the *relative* differences in society costs when different allocation strategies are used.

Distribution to Non-Covered Entities

Non-covered entities may receive free allowances (sometimes referred to as "set-asides") or a percentage of the auction revenues. For example, Congress may decide to allot allowance value to electricity consumers, particularly those in low-income households. The rationale for this distribution policy is that specific subsets of society are expected to bear a disproportionate percentage of the costs of a cap-and-trade program.

Whether covered sources receive allowances at no cost or purchase them through an auction, economic principles predict that covered sources would pass along their opportunity costs or purchase costs, respectively, in the same manner as an actual expense: e.g., installing more efficient technology or switching to more expensive (but less carbon-intensive) fuels. Covered sources have demonstrated this behavior in two cap-and-trade programs, in which the vast majority of allowances was provided at no cost: the European Union's Emission Trading System (EU-ETS) and the U.S. sulfur dioxide emissions trading program.[86]

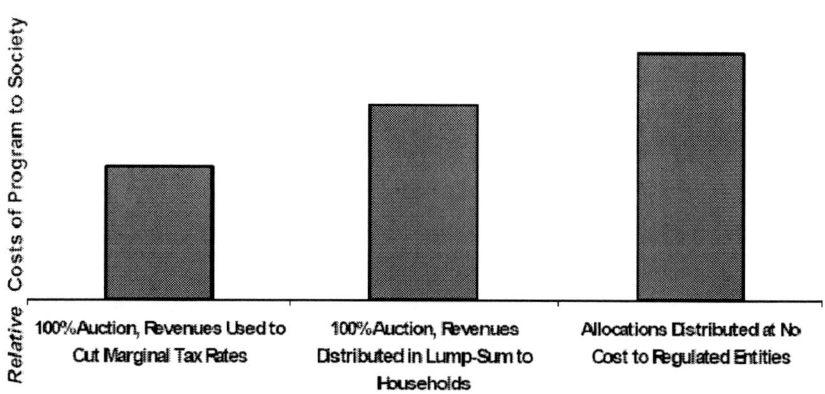

Source: Prepared by CRS with data from Lawrence H. Goulder, *Mitigating the Adverse Impacts of CO_2 Abatement Policies on Energy-Intensive Industries* (2002), Resources for the Future Discussion Paper.

Figure 4. Relative Differences in Cap-and-Trade Program's Cost to Society with Different Emission Allocation Strategies.

Because of cost pass-through, consumers, particularly households, are ultimately expected to bear the majority of the costs associated with a cap-and-trade program. Figure 5 illustrates the *relative* distribution of costs to different groups in a cap-andtrade program.[87] The figure is based on a cap-and-trade scenario that would auction 100% of the emission allowances to fossil fuel producers (often referred to as "upstream" sources). Households and businesses experience the vast majority (89%) of the costs. Moreover, the household percentage is potentially understated, because many businesses would likely pass through some of their increased energy/electricity costs in the form of higher prices for their goods and services.[88]

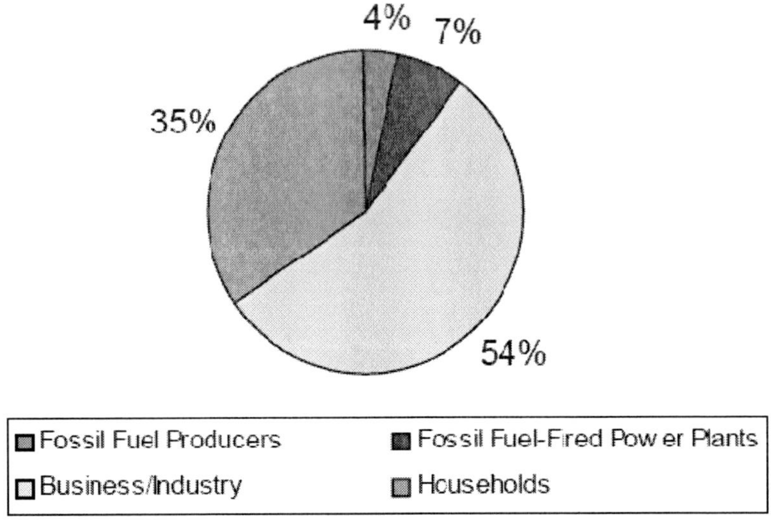

Source: Prepared by CRS based on the data from the National Commission on Energy Policy, *Allocating Allowances in a Greenhouse Gas Trading System* (2007).

Note: The percentages above do not account for any offsetting income from allowance allocation. The figure illustrates the relative distributions that would occur if allowances were auctioned to fossil fuel producers ("upstream"), without recycling the revenues.

Figure 5. Relative Distribution of Costs Using Upstream Auction (*without Revenue Redistribution*).

The figure is instructive for the allocation debate, because it shows a starting point for cost distribution. However, the cost percentages depicted in Figure 5 do not account for distribution of auction revenues. In a real cap-and-trade system that employs an auction, auction revenues would be used to support specific objectives, or allotted to various parties.

Distribution to Support Specific Objectives

Policymakers may also consider distributing a percentage of the allowances or auction revenues to support a range of objectives, including:

- Technology development: Promotion of emission mitigation technology is widely recognized as a vital step towards making substantial GHG emission reductions.[89] S. 2191 would distribute the majority of its auction revenues (approximately 52%) to promote energy-related technologies: low-carbon energy sources, carbon capture and storage (CCS),[90] and cellulosic biofuels.

- Energy efficiency: Improvements in energy efficiency could make considerable contributions in achieving GHG emission reductions. Although energy efficiency may involve technology advancements, new technologies must be used in order to realize the efficiency gains. In some cases, parties may need incentives beyond the efficiency gains to induce behavioral changes. For example, allowance value could be distributed to support efficiency gains at places — residences, commercial buildings — that are unlikely to be covered by an emissions cap.
- Biological sequestration: Trees, plants, and soils sequester carbon, removing it from the earth's atmosphere. Allowance values could be allotted to provide financial incentives for landowners to engage in activity — e.g., conservation tillage, reforestation — that would increase sequestration on their land. Although some of these actions may qualify as offsets (if allowed in a cap-and-trade program), some activities may need additional incentives or be unable to satisfy the offset approval process.[91]
- Adaptation efforts: Some level of global warming (and associated effects) will occur regardless of emission reduction efforts taken today, because previous and current GHG emissions will have long term climate impacts. Therefore, some contend that investment (e.g., allowance value) should focus on preparing communities to adapt to the effects of a changing climate.
- Deficit reduction: Another proposal for the use of allowances or auction revenue is to address a budget deficit that may result from a cap-and-trade program. A portion of the allowances or auction proceeds could be set aside to offset projected revenue shortfalls. This option is sometimes described as making the program "revenue-neutral." In addition, Congress may consider using allowance value to offset shortfalls beyond those that are related to the cap-and-trade program.

Policy Considerations

The distribution of allowance value would present policymakers with a series of trade-offs. The primary options for applying the allowance value involve (1) minimizing the overall costs of the cap-and-trade program imposed on society; (2) alleviating the disproportionate costs borne by subgroups in society; and (3) providing funding to support other policy objectives, such as technological development. Intertwined among these options is a trade-off between regressive or progressive impacts.

Reduce Costs, Alleviate Burdens, or Promote Technology

Economic studies indicate that using auction revenues (i.e., revenue recycling) to offset reductions in distortionary taxes (labor, income) would be the most efficient use of the revenues and yield the greatest benefit to society as a whole. These studies show that if the revenues are used for other purposes, economic efficiency would suffer and the overall cost of the program would be higher (Figure 4).[92]

The opportunity to substantially lower overall costs is a unique attribute of the auction allocation strategy. Some argue that using auction revenues to offset tax reductions is an unlikely outcome.[93] In general, when the government obtains new revenue, it tends to fund new or existing programs rather than reduce existing taxes.[94] Indeed, none of the cap-and-trade bills in Congress has proposed to use revenues in this manner.[95]

Moreover, the most efficient manner of revenue recycling would generally spread the cost reductions throughout the economy, while certain groups would be expected to bear a disproportionate percentage of the costs of a cap-and-trade program. In particular, industries that cannot pass along their increased costs (for competitive reasons or otherwise) would face higher costs compared to other economic sectors. Ultimately, consumers are expected to absorb the vast majority of the program costs. As discussed above, policymakers may consider compensating — through free allowances or auction revenues — specific industries and/or providing assistance to households that would face higher energy (e.g., electricity and gasoline) prices. Such compensation would forgo the opportunity to reduce the overall cost of the emission program.

Using allowance value for other purposes — e.g., technology development — would impart a trade-off. While such allotments would limit the allowances available for the purposes described above — reduced overall costs or relief for disproportionate impacts — new technologies, in particular, could have a considerable impact on the costs of an emissions reduction program. This is particularly the case if technologies can be commercialized ahead of their projected schedules[96] or if unanticipated low- or zero-carbon alternatives can be developed.

The amount of funding allotted to technology development under S. 2191 would represent a substantial increase, compared to current federal funding for emission reduction and low-carbon technologies. Although allocating allowance value to technological development could yield considerable gain, the return in terms of additional investment to promote technology improvements remains uncertain.[97] In a recent study, the author found:

> While academics have extensively studied how revenues from auctions can be used to make the tax system more efficient, they have not done much research on examining the impact of using such revenues for retiring debt or increasing government investment.

In short, if allowance value is used to increase funding to existing programs or provide funding for new programs, the return on investment is uncertain. For example, some programs may be operating at full capacity, and additional funding may not provide comparable impacts. Moreover, the marketplace already provides incentive for technological change, and a cap-and-trade program would increase the incentive to develop low-carbon or zero-carbon energy alternatives.

Regressive or Progressive Economic Effects

Another trade-off for policymakers is whether the emission allocation strategy would produce regressive or progressive economic results.[98] The Congressional Budget Office (CBO) analyzed the distributional effects of a cap-and-trade program under four different scenarios:

- Allowances auctioned and auction revenues distributed to households in a lump-sum payment;
- Allowances provided to covered sources at no cost;
- Allowances auctioned and revenues are used to cut corporate income taxes; and
- Allowances auctioned and revenues used to cut payroll income taxes.

The results are presented in Figure 6. As the figure indicates, an auction that distributed the revenues to households in a lump-sum payment would yield the most progressive results. In contrast, the most regressive result occurs when allowances are provided to covered sources at no cost. The two other options that would cut tax rates would also produce regressive results.

However, these two strategies would substantially lower the overall cost of the cap-and-trade program, a result not captured by Figure 6.

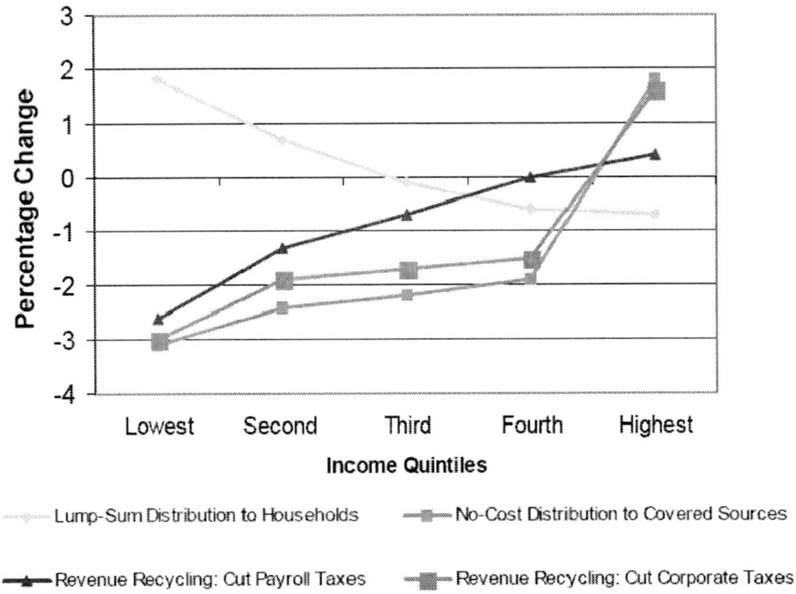

Source: Prepared by CRS with data from the CBO: the no-cost distribution scenario data are from CBO, *Who Gains and Who Pays Under Carbon-Allowance Trading? The Distributional Effects of Alternative Policy Designs* (2000); the other three scenario data are from CBO, *Trade-Offs in Allocating Allowances for CO_2 Emissions* (2007), Economic and Budget Issue Brief.

Figure 6. Comparison of After-Tax Household Income Changes (by Quintile) Imposed by Emissions Cap, Using Different Emission Allocation and Revenue Distribution Strategies.

APPENDIX A. WHAT IS A CAP-AND-TRADE SYSTEM?

A cap-and-trade system would create an overall limit (i.e., a cap) on GHG emissions from the emission sources covered by the program. Cap-and-trade programs can vary by the sources covered. The covered sources are likely to include major emitting sectors (e.g., power plants and carbon-intensive industries), fuel producers/processors (e.g., coal mines or petroleum refineries), or some combination of both.

The emissions cap is partitioned into emission allowances. Typically, one emission allowance represents the authority to emit one (metric) ton of carbon dioxide-equivalent (tCO_2-e). The "equivalent" is necessary, because GHGs other than CO_2 — methane, nitrous oxide, sulfur hexafluoride, hydrofluorocarbons, and perfluorocarbons — vary in their global warming potential (GWP).[99] Thus, GHG emissions are presented in a standard form of measure (CO_2-e).

In general, policymakers may decide to distribute the emission allowances to covered entities at no cost (based on, for example, previous years' emissions), sell the allowances through an auction, or use some combination of these strategies. This report examines issues associated with these allocation options.

Covered entities that face relatively low emission-reduction costs would have an incentive to make reductions beyond what is required, because these further reductions could be sold (i.e., traded) as emission credits to entities that face higher emission-reduction costs. Other mechanisms, such as banking or offsets, may be included to increase the flexibility of the program.

At the end of each established compliance period (e.g., a calendar year), covered sources would be required to surrender emission allowances to cover the number of tons emitted. If a source did not have enough allowances to cover its emissions, the source would be subject to penalties.

For more information, see U.S. Environmental Protection Agency (EPA), Office of Air and Radiation, *Tools of the Trade: A Guide To Designing and Operating a Cap and Trade Program For Pollution Control* (2003); CRS Report RL33799, *Climate Change: Design Approaches for a Greenhouse Gas Reduction Program*, by Larry Parker.

APPENDIX B. ALLOWANCE ALLOCATION STRATEGY UNDER S. 2191 (AS REPORTED)

Table 2. Emission Allowance Allocation under S. 2191

	2012	**2015**	**2020**	**2025**	**2030**	**2035**	**2040**	**2045**	**2050**
	Allowances Sold through an Auction								
Early Auction	5%								
Auction	21.5%	29.5%	36.5%	48.5%	62.8%	69.5%	69.5%	69.5%	69.5%
	Allowances Distributed at No Cost								
	2012	**2015**	**2020**	**2025**	**2030**	**2035**	**2040**	**2045**	**2050**
States									

Table 2. (Continued)

	2012	2015	2020	2025	2030	2035	2040	2045	2050
Energy Savings	2%	2%	2%	2%	2%	2%	2%	2%	2%
Building Efficiency	1%	1%	1%	1%	1%	1%	1%	1%	1%
Programs that Exceed Fed. Targets	2%	2%	2%	2%	2%	2%	2%	2%	2%
General Allocation - by LIHEAP Share	1.5%	1.5%	1.5%	1.5%	1.5%	1.5%	1.5%	1.5%	1.5%
Population Share	1.5%	1.5%	1.5%	1.5%	1.5%	1.5%	1.5%	1.5%	1.5%
Fossil Production CO_2 Share	1.5%	1.5%	1.5%	1.5%	1.5%	1.5%	1.5%	1.5%	1.5%
Mass Transit	1%	1%	1%	1%	1%	1%	1%	1%	1%
	Transition Assistance								
Fossil Fueled Electric Plants	19%	19%	16%	10%	1%				
Rural Electric Cooperatives	1%	1%	1%	1%	1%				
Pilot Program for VA and MT	0.2%	0.2%	0.2%	0.2%					
Energy-Intensive Manufacturing	10%	10%	8%	4%					
Petroleum Production/Importers	2%	2%	2%	1.00%	0.25%				
HFC Producers/Importers	2%	2%	2%	1.00%	0.25%				
	Other Purposes or Recipients								
Early Action	5%	2%							
Tribal Communities	0.5%	0.5%	0.5%	0.5%	0.5%	0.5%	0.5%	0.5%	0.5%
Low/Middle-Income Electricity Consumers	9%	9%	9%	9%	9%	9%	9%	9%	9%
Low/Middle-Income Natural Gas Consumers	2%	2%	2%	2%	2%	2%	2%	2%	2%
Carbon Capture and Sequestration	4%	4%	4%	4%	4%				
Domestic Agriculture and Forestry	5%	5%	5%	5%	5%	5%	5%	5%	5%
International Forest Protection	2.5%	2.5%	2.5%	2.5%	2.5%	2.5%	2.5%	2.5%	2.5%
Landfill and Coal Mine CH_4 Reduction	3%	3%	3%	3%	3%	3%	3%	3%	3%

Source: Prepared by CRS.

Table 3. Auction Revenue Distribution under S. 2191

Off-the-Top Allocation of Auction Revenues (in $1,000s)									
	2012	2015	2020	2025	2030	2035	2040	2045	2050
BLM Emergency Firefighting Fund	$150	$150	$150	$150	$150	$150	$150	$150	$150
Forest Service Emergency Firefighting Fund	$430	$430	$430	$430	$430	$430	$430	$430	$430

	2012	2015	2020	2025	2030	2035	2040	2045	2050
Climate Security Act Management Fund	$1,071	$1,211	$1,393	$1,586	$1,776	$1,950	$2,086	$2,149	$2,092
Percentage Allocation of Remaining Revenues									
	2012	2015	2020	2025	2030	2035	2040	2045	2050
Technology Deployment									
Zero- or Low- Carbon Energy Technology	16.6%	16.6%	16.6%	16.6%	16.6%	16.6%	16.6%	16.6%	16.6%
Advanced Coal and Sequestration Technology	13.0%	13.0%	13.0%	13.0%	13.0%	13.0%	13.0%	13.0%	13.0%
Fuel from Cellulosic Biomass	3.1%	3.1%	3.1%	3.1%	3.1%	3.1%	3.1%	3.1%	3.1%
Advanced Technology Vehicles Manufacturing	6.2%	6.2%	6.2%	6.2%	6.2%	6.2%	6.2%	6.2%	6.2%
Sustainable Energy Program	13.0%	13.0%	13.0%	13.0%	13.0%	13.0%	13.0%	13.0%	13.0%
TOTAL	52.0%	52.0%	52.0%	52.0%	52.0%	52.0%	52.0%	52.0%	52.0%
Energy Assistance Fund									
LIHEAP	9.0%	9.0%	9.0%	9.0%	9.0%	9.0%	9.0%	9.0%	9.0%
Weatherization	4.5%	4.5%	4.5%	4.5%	4.5%	4.5%	4.5%	4.5%	4.5%
Rural Energy Assistance	4.5%	4.5%	4.5%	4.5%	4.5%	4.5%	4.5%	4.5%	4.5%
TOTAL	18.0%	18.0%	18.0%	18.0%	18.0%	18.0%	18.0%	18.0%	18.0%
Climate Change Worker Training Fund									
DOE University Programs	1.3%	1.3%	1.3%	1.3%	1.3%	1.3%	1.3%	1.3%	1.3%
TOTAL	5.0%	5.0%	5.0%	5.0%	5.0%	5.0%	5.0%	5.0%	5.0%
Adaptation Fund									
DOI - Wildlife Conservation and Restoration	6.3%	6.3%	6.3%	6.3%	6.3%	6.3%	6.3%	6.3%	6.3%
DOI - Adaptation Activities	3.4%	3.4%	3.4%	3.4%	3.4%	3.4%	3.4%	3.4%	3.4%
DOI - Cooperative Grant Programs	0.9%	0.9%	0.9%	0.9%	0.9%	0.9%	0.9%	0.9%	0.9%
DOI - Tribal Wildlife Grants	0.2%	0.2%	0.2%	0.2%	0.2%	0.2%	0.2%	0.2%	0.2%
Land and Water Conservation Fund									
DOI LWCF Sec. 6 Grants	*0.3%*	*0.3%*	*0.3%*	*0.3%*	*0.3%*	*0.3%*	*0.3%*	*0.3%*	*0.3%*
DOI LWCF Sec. 7 Acquisitions	*0.6%*	*0.6%*	*0.6%*	*0.6%*	*0.6%*	*0.6%*	*0.6%*	*0.6%*	*0.6%*
USDA Forest Legacy Program Sec. 7 Acquisitions	*0.3%*	*0.3%*	*0.3%*	*0.3%*	*0.3%*	*0.3%*	*0.3%*	*0.3%*	*0.3%*
USDA LWCF Sec. 7 Acquisitions	*0.6%*	*0.6%*	*0.6%*	*0.6%*	*0.6%*	*0.6%*	*0.6%*	*0.6%*	*0.6%*
SUBTOTAL	1.8%	1.8%	1.8%	1.8%	1.8%	1.8%	1.8%	1.8%	1.8%
Forest Service Adaptation Activities	0.9%	0.9%	0.9%	0.9%	0.9%	0.9%	0.9%	0.9%	0.9%
EPA Adaptation Activities	0.9%	0.9%	0.9%	0.9%	0.9%	0.9%	0.9%	0.9%	0.9%
Army Corps of Engineers Adaptation Activities	1.8%	1.8%	1.8%	1.8%	1.8%	1.8%	1.8%	1.8%	1.8%
Department of Commerce Adaptation Activities	1.8%	1.8%	1.8%	1.8%	1.8%	1.8%	1.8%	1.8%	1.8%
TOTAL	18.0%	18.0%	18.0%	18.0%	18.0%	18.0%	18.0%	18.0%	18.0%

Table 3. (Continued)

	2012	2015	2020	2025	2030	2035	2040	2045	2050
Energy Independence Acceleration Fund	2.0%	2.0%	2.0%	2.0%	2.0%	2.0%	2.0%	2.0%	2.0%
Climate Change and National Security Fund	5.0%	5.0%	5.0%	5.0%	5.0%	5.0%	5.0%	5.0%	5.0%

Source: Prepared by CRS.

Notes: CRS estimates of off-the-top revenues based on allowance price projections from the EPA/ADAGE-TECH scenario in U.S. EPA, *EPA Analysis of the Lieberman-Warner Climate Security Act of 2008* (2008). Higher allowance price estimates would lead to higher auction proceeds. CRS estimates of firefighting fund requirements based on historic data. Estimate of administration cost ("CSA Management Fund") based on EPA's estimate of 1% of total allowance value.

End Notes

[1] The Intergovernmental Panel on Climate Change (IPCC) concluded in 2007 that "most of the observed increase in globally averaged temperatures since the mid-20th century is *very likely* due to the observed increase in anthropogenic greenhouse gas concentrations" (emphasis added). IPCC Working Group I, *Climate Change 2007: The Physical Basis* (Cambridge, UK: Cambridge University Press, 2007). See CRS Report RL34266, *Climate Change: Science Update 2007*, by Jane Leggett.

[2] Senate Amendment No. 866 to H.R. 6 (109th Congress), passed by voice vote June 22, 2005. A motion to table the amendment was rejected by a roll call vote (44 - 53).

[3] The policy alternative to a market-based approach would likely require specific emission limits or particular technological controls for specific emission sources.

[4] See CRS Report RL34067, *Climate Change Legislation in the 110th Congress*, by Jonathan L. Ramseur and Brent Yacobucci.

[5] Another market-based approach would entail a carbon tax. Some of the cap-and-trade proposals include elements (e.g., safety-valve) that are akin to a carbon tax. These proposals are often described as hybrid approaches. See CRS Report RL33846, *Greenhouse Gas Reduction: Cap-and-Trade Bills in the 110th Congress*, by Larry Parker and Brent D. Yacobucci.

[6] U.S. Environmental Protection Agency (EPA), Office of Air and Radiation, *Tools of the Trade: A Guide To Designing and Operating a Cap and Trade Program For Pollution Control* (2003).-B-03-002.

[7] There are two noteworthy exceptions: electric utilities operating in a price-regulated market (discussed below) and facilities that receive allowances based on an output-based distribution system. Robert Stavins, *A U.S. Cap-and-Trade System to Address Global Climate Change* (2007), The Hamilton Project, Brookings Institution.

[8] Congressional Budget Office, *Trade-Offs in Allocating Allowances for CO_2 Emissions* (2007), Economic and Budget Issue Brief.

[9] Like there is no free lunch, free allowances are not really free. However, this report uses the phrase "distribution at no cost" and "free allowances" interchangeably.

[10] In economics parlance, this is referred to as a firm's "opportunity cost."

[11] See National Commission on Energy Policy, *Allocating Allowances in a Greenhouse Gas Trading System* (2007); Congressional Budget Office, *Trade-Offs in Allocating Allowances for CO_2 Emissions* (2007), Economic and Budget Issue Brief; Dallas Burtraw, *Cap, Auction, and Trade: Auctions Revenue Recycling under Carbon Cap and Trade* (2008), Testimony Prepared for the House Select Committee on Energy Independence and Global Warming.

[12] The 10 states participating in the Regional Greenhouse Gas Initiative (RGGI) have agreed to auction at least 25% of their allowances, and several of the RGGI states intend to auction almost 100% of their allotments. See CRS Report RL33812, *Climate Change: Action by States To Address Greenhouse Gas Emissions*, by Jonathan L. Ramseur.

[13] For a comparison of the cap-and-trade bills see CRS Report RL33846, *Greenhouse Gas Reduction: Cap-and-Trade Bills in the 110th Congress*, by Larry Parker and Brent D. Yacobucci.

[14] The logistics of establishing and running an auction are more complicated. This issue is discussed below.

[15] U.S. Environmental Protection Agency (EPA), Office of Air and Radiation, *Tools of the Trade: A Guide To Designing and Operating a Cap and Trade Program For Pollution Control* (2003).

[16] This behavior is described as "rent-seeking" in economic contexts. Dallas Burtraw, Prepared Testimony before the House Select Committee on Energy Independence and Global Warming, January 23, 2008.

[17] This cost savings estimate is based on an analysis that simulated a cap-and-trade program with a 22% emission reduction between 2000 and 2080. Lawrence H. Goulder, *Mitigating the Adverse Impacts of CO_2 Abatement Policies on Energy-Intensive Industries* (2002), Resources for the Future Discussion Paper.

[18] See e.g., Goulder (2002); Anne E. Smith and Martin T. Ross, *Allowance Allocation: Who Wins and Loses Under a Carbon Dioxide Control Program?* (2002), Charles River Associates; Dallas Burtraw, et al., *The Effect of Allowance Allocation on the Cost of Carbon Emission Trading* (2001), Resources for the Future.

[19] A carbon tax system, which is not within the purview of this report, could achieve the same result, if carbon tax proceeds were applied in a similar manner.

[20] See e.g., Robert W. Hahn, *Greenhouse Gas Auctions and Taxes: Some Practical Considerations* (2008), AEI Center for Regulatory and Market Studies; Robert Stavins, *A U.S. Cap-and-Trade System to Address Global Climate Change* (2007), The Hamilton Project, Brookings Institution.

[21] Charles Holt *et al.*, *Auction Design for Selling CO_2 Emission Allowances Under the Regional Greenhouse Gas Initiative* (2007), prepared for RGGI Working Group staff..

[22] Holt *et al.* (2008).

[23] Note that this value will fluctuate daily with changes in fuel prices and energy demands.

[24] Holt *et al.* (2008).

[25] More information on EPA's SO2 emission trading auction is at [http://www.epa.gov/airmarkets/trading/factsheet-auction.html#how].

[26] Ken Macken (Ireland Environmental Protection Agency), Presentation for RGGI Auction Workshop, March 2006, at [http://www.rggi.org/documents.htm].

[27] Holt *et al.* (2008).

[28] For more information about Virginia's auction, see William Shobe, Presentation for RGGI Auction Workshop, March 2006, at [http://www.rggi.org/documents.htm].

[29] Holt *et al.* (2008).

[30] Ibid.

[31] Ibid.

[32] Although EPA annually auctions a small percentage of the allowances in its sulfur dioxide (SO_2) cap-and-trade program (pursuant to Title IV of the 1990 Clean Air Act Amendments), most of the allowances are provided at no cost to emission sources, based on their historical emissions. See CRS Report RL34235, *Air Pollution as a Commodity: Regulation of the Sulfur Dioxide Allowance Market*, by Larry Parker and Mark Jickling.

[33] The European Union's (EU) Emissions Trading Scheme (ETS), a CO_2 cap-and-trade program that applies to power plants and certain carbon-intensive industries, allowed countries (between 2005-2008) to auction up to 5% of allowance allocations. Only 4 of 25 countries used auctions at all, and only Denmark auctioned the full 5%. See CRS Report RL34150, *Climate Change: The EU Emissions Trading Scheme (ETS) Enters Kyoto Compliance Phase*, by Larry Parker.

[34] See e.g., Cameron Hepburn, *et al.*, "Auctioning of EU ETS phase II allowances: how and why?" (2006) *Climate Policy* 6(1): 137-160.

[35] This may vary by the product sold and the level of price increase. Some products may display a relatively inelastic price/demand relationship. This discussion is beyond the scope of this report.

[36] For industry-specific estimates, see e.g., U.S. EPA, *EPA Analysis of the Lieberman-Warner Climate Security Act of 2008* (2008).

[37] National Commission on Energy Policy, *Allocating Allowances in a Greenhouse Gas Trading System* (2007), p. 10.

[38] U.S. EPA, *Inventory of U.S. Greenhouse Gas Emissions and Sinks: 1990-2006* (2008), Table ES-7, at [http://epa.gov/climatechange/emissions/usinventoryreport.html].

[39] See e.g., National Commission on Energy Policy, *Allocating Allowances in a Greenhouse Gas Trading System* (2007).

[40] Dallas Burtraw, Prepared Testimony before the House Select Committee on Energy Independence and Global Warming, January 23, 2008.

[41] The above percentages are calculated by CRS with data from the Energy Information Administration's 906/920 database, available at [http://www.eia.doe.gov/cneaf/electricity/page/eia906_920.html].

[42] These figures are from Dallas Burtraw, Prepared Testimony before the House Select Committee on Energy Independence and Global Warming, January 23, 2008. This testimony cites the author's more in-depth study: Dallas Burtraw and Karen Palmer, *Compensation Rules for Climate Policy in the Electricity Sector* (2007), Resources for the Future Discussion Paper.

[43] These figures are provided for illustrative purposes only.

[44] For example, electricity in some states or regions may use higher percentages of coal or renewable energy than other areas. See CRS Report RL34272, *State Greenhouse Gas Emissions: Comparison and Analysis*, by Jonathan L. Ramseur.

[45] Calculated by CRS with data from the Energy Information Administration's 906/920 database, available at [http://www.eia.doe.gov/cneaf/electricity/page/eia906_920.html].

[46] National Commission on Energy Policy, *Allocating Allowances in a Greenhouse Gas Trading System* (2007).

[47] See Lawrence H. Goulder, *Mitigating the Adverse Impacts of CO_2 Abatement Policies on Energy-Intensive Industries* (2002), Resources for the Future Discussion Paper; Anne E. Smith and Martin T. Ross, *Allowance Allocation: Who Wins and Loses Under a Carbon Dioxide Control Program?* (2002), Charles River Associates.

[48] Compared to other emissions trading programs — namely, the U.S. SO_2 emissions (acid rain) program — the ratio of allowance value to compliance costs is expected to be much greater.

[49] As discussed above, price-regulated electric utilities represent a critical exception.

[50] This is due to the opportunity costs that an entity would face if it used (i.e., surrendered to the implementing agency for compliance) its allowances. Instead of using the allowance, the entity could have sold it for its market value.

[51] However, higher prices could reduce consumer demand and potentially lower profits (discussed below).

[52] U.S. Congress, Senate Committee on Energy and Natural Resources, *Design Elements of a Mandatory Market-Based Greenhouse Gas Regulatory System*, Chairman and Ranking Member Statement: Climate Change Conference (2006), 109th Congress.

[53] Goulder (2002) simulated a U.S. cap-and-trade program that would require a 23% emission reduction. In the model's scenario that distributed allowances at no cost to energy producers (coal, oil, natural gas), the coal sector profits increased by 155% after two years of the program.

[54] Converted using exchange rate of 1.25 (average rate in 2005), provided by [http://www.oanda.com].

[55] IPA Energy Consulting, *Implications of the EU Emissions Trading Scheme for the UK Power Generation Sector* (2005), Prepared for the United Kingdom Department of Trade and Industry.

[56] Converted using exchange rate of 1.25 (average rate in 2005), provided by [http://www.oanda.com].

[57] Jos Sijm *et al.*, "CO_2 Cost Pass-Through and Windfall Profits in the Power Sector," *Climate Policy* 6 (2006): 49-72.

[58] See CRS Report RL34150, *Climate Change: The EU Emissions Trading Scheme (ETS) Enters Kyoto Compliance Phase*, by Larry Parker.

[59] Using exchange rate of 1.5, provided by [http://www.oanda.com].

[60] Point Carbon Advisory Services, *EU ETS Phase II — The Potential and Scale of Windfall Profits in the Power Sector* (2008), Prepared for World Wildlife Fund.

[61] Proposal for a Directive of the European Parliament and of the Council amending Directive 2003/87/EC so as to improve and extend the greenhouse gas emission allowance trading system of the Community (January 23, 2008), at [http://ec.europa.eu/environment/climat /emission/ets_post2012_en.htm].

[62] In some cases, where the firm might shift these operations to a foreign, unregulated country, such incentives might make sense. Raymond J. Kopp, 2007, *Allowance Allocation*, Resources for the Future.

[63] Markus Ahman *et al.*, "A Ten-Year Rule to Guide the Allocation of EU Emission Allowances," *Energy Policy* 35 (2007):1718-1730.

[64] However, some objectives, if met — namely, technology advancement — may also reduce the overall costs of the program. This is discussed below.

[65] See CRS Report RL34436, *The Role of Offsets in a Greenhouse Gas Emissions Cap-and-Trade Program: Potential Benefits and Concerns*, by Jonathan L. Ramseur.

[66] These costs may be described with different terms in different publications: e.g., program costs, mitigation costs, or economic costs.

[67] For the more stringent cap-and-trade proposals in the 110th Congress, this threshold would not be reached until approximately 2040. See CRS Report RL33846, *Greenhouse Gas Reduction: Cap-and-Trade Bills in the 110th Congress*, by Larry Parker and Brent Yacobucci.

[68] Raymond J. Kopp, *Allowance Allocation* (2007), Resources for the Future.

[69] The option to bank emission allowances would likely alter this calculus. In anticipation that allowance prices will increase in subsequent years, covered sources would likely purchase more allowances than are needed for compliance purposes in the early years of the program.

[70] Typically, "economy-wide" proposals would cover the vast majority of the nation's GHG emissions by capping electricity generation, carbon-intensive industries, and the transportation sector. Depending on the design of the program, some sectors (e.g., agricultural or residential) may be excluded from the cap.

[71] U.S. EPA, *EPA Analysis of the Lieberman-Warner Climate Security Act of 2008* (2008).

[72] U.S. Internal Revenue Service, *Internal Revenue Service Data Book 2007*, Publication 55B, issued March 2008.

[73] Lawrence H. Goulder, *Mitigating the Adverse Impacts of CO_2 Abatement Policies on Energy-Intensive Industries* (2002), Resources for the Future Discussion Paper.

[74] Dallas Burtraw and Karen Palmer, *Compensation Rules for Climate Policy in the Electricity Sector* (2007), Resources for the Future Discussion Paper.

[75] The authors found a range of 9% to 21%, but noted that the 21% was consistent with the scenario presented in Goulder's study. Anne Smith *et al.*, *Implications of trading Implementation Design for Equity-Efficiency Trade-Offs in Carbon Permit Allocations* (2002), Charles River Associates.

[76] Dallas Burtraw and Karen Palmer, *Compensation Rules for Climate Policy in the Electricity Sector* (2007), Resources for the Future Discussion Paper.

[77] Section 3901 (reported May 20, 2008).

[78] In addition, note that S. 2191 would not provide allowances directly to the coal industry, although petroleum producers/importers would receive a small percentage (2%).

[79] The rationale for this policy is that taxes on desirable activities create market distortions, discouraging increased levels of desirable actions. Assuming the same amount of revenue could be collected, economic policy would favor placing the tax on activity that is generally considered undesirable. Further discussion is beyond the scope of this report.

[80] Ian Parry, "Fiscal Interactions and the Case for Carbon Taxes over Grandfathered Carbon Permits," in *Climate Change Policy* (Dieter Helm, editor), Oxford University Press (2005); Intergovernmental Panel on Climate Change Working Group III, *Climate Change 2001: Mitigation*, (Cambridge, UK: Cambridge University Press, 2001) Chapter 8.

[81] This is referred to as the "tax-interaction effect" in economic literature. See e.g., Ian Parry, "Fiscal Interactions and the Case for Carbon Taxes over Grandfathered Carbon Permits," in *Climate Change Policy* (Dieter Helm, editor), Oxford University Press (2005)

[82] There is some evidence that the double-dividend benefits are stronger in Europe than in the United States, because the former has a more stringent tax system. Intergovernmental Panel on Climate Change Working Group III, *Climate Change 2001: Mitigation*, (Cambridge, UK: Cambridge University Press, 2001) Chapter 8, citing several studies.

[83] See Lawrence H. Goulder, *Mitigating the Adverse Impacts of C)2 Abatement Policies on Energy-Intensive Industries* (2002), Resources for the Future Discussion Paper; and Anne E. Smith and Martin T. Ross, *Allowance Allocation: Who Wins and Loses Under a Carbon Dioxide Control Program?* (2002), Charles River Associates; Dallas Burtraw, et al., *The Effect of Allowance Allocation on the Cost of Carbon Emission Trading* (2001), Resources for the Future.

[84] See CRS Report RL34489, *Climate Change: Costs and Benefits of S. 2191*, by Larry Parker and Brent Yacobucci.

[85] Goulder (2002). This analysis simulated a cap-and-trade program with a 22% emission reduction between 2000 and 2080.

[86] Congressional Budget Office, *Trade-Offs in Allocating Allowances for CO_2 Emissions* (2007), Economic and Budget Issue Brief.

[87] This figure is based on a National Commission on Energy Policy (NCEP) proposal that would lead to a relatively modest reduction in GHG emissions compared to those required under current proposals in the 110th Congress. Thus, this figure is illustrative and only useful for comparing relative differences.

[88] National Commission on Energy Policy, *Allocating Allowances in a Greenhouse Gas Trading System* (2007).

[89] See e.g., CBO, *Evaluating the Role of Prices and R&D in Reducing Carbon Dioxide Emissions*, (2006).

[90] See CRS Report RL33801, *Carbon Capture and Sequestration (CCS)*, by Peter Folger.

[91] See CRS Report RL34436, *The Role of Offsets in a Greenhouse Gas Emissions Cap-and-Trade Program: Potential Benefits and Concerns*, by Jonathan L. Ramseur.

[92] However, if additional funding for technology advancement leads to faster development of low-carbon technologies, the cost of the program may decrease (discussed below).

[93] Robert W. Hahn, *Greenhouse Gas Auctions and Taxes: Some Practical Considerations* (2008), AEI Center for Regulatory and Market Studies; Robert Stavins, *A U.S. Cap-andTrade System to Address Global Climate Change* (2007), The Hamilton Project, Brookings Institution.

[94] Ibid, citing Gary Becker and Casey Mulligan, "Deadweight Costs and the Size of the Government (2003)," *Journal of Law and Economics* 46(2).

[95] Another obstacle to this approach may concern Congressional committee jurisdiction issues. A cap-and-trade proposal that would restructure elements of the tax code may overlap with multiple committee jurisdictions.

[96] For example, the cost models for S. 2191 make assumptions about the availability of CCS. When this technology actually goes online, it will likely have a substantial impact on the cost of a cap-and-trade program.

[97] For more discussion of recent modeling results regarding technological development, see CRS Report RL34489, *Climate Change: Costs and Benefits of S. 2191*, by Larry Parker and Brent Yacobucci.

[98] Analogous to a regressive tax policy, a regressive allocation strategy would disproportionately impact lower income households. In contrast, in a progressive distribution scheme the ratio of cost to income would increase as income rises.

[99] GWPs are used to compare gases to carbon dioxide, which has a GWP of 1. For example, methane's GWP is 25, and is thus 25 times more potent a GHG than CO_2. GWPs are typically based on estimates provided by the Intergovernmental Panel on Climate Change (IPCC).

In: The Regional Greenhouse Gas Initiative
Editor: Serge Chauvin
ISBN: 978-1-62808-483-2

Chapter 3

THE ROLE OF OFFSETS IN A GREENHOUSE GAS EMISSIONS CAP-AND-TRADE PROGRAM: POTENTIAL BENEFITS AND CONCERNS*

Jonathan L. Ramseur

SUMMARY

If Congress establishes a greenhouse gas (GHG) emissions reduction program (e.g., cap-and-trade system), the treatment of GHG emission offsets would likely be a critical design element. If allowed as part of an emissions program, offsets could provide cost savings and other benefits. However, offsets have generated concern.

An offset is a measurable reduction, avoidance, or sequestration of GHG emissions from a source not covered by an emission reduction program. If allowed, offset projects could generate "emission credits," which could be used by a regulated entity (e.g., power plant) to comply with its reduction requirement. Offsets could include various activities:

- agriculture or forestry projects: e.g., conservation tillage or planting trees on previously non-forested lands;
- renewable energy projects: e.g., wind farms;

---

* This is an edited, reformatted and augmented version of Congressional Research Service, Publication No. RL34436, dated May 18, 2009.

- energy efficiency projects: e.g., equipment upgrades;
- non-CO_2 emissions reduction projects: e.g., methane from landfills.

Including offsets would likely make an emissions program more cost-effective by (1) providing an incentive for non-regulated sources to generate emission reductions and (2) expanding emission compliance opportunities for regulated entities. Some offset projects may provide other benefits, such as improvements in air or water quality. In addition, the offset market may create new economic opportunities and spur innovation as parties seek new methods of generating offsets.

The main concern with offset projects is whether or not they represent real emission reductions. For offsets to be credible, a ton of CO_2-equivalent emissions from an offset project should equate to a ton reduced from a covered emission source, such as a smokestack or exhaust pipe. This objective presents challenges because many offsets are difficult to measure. If illegitimate offset credits flow into an emissions trading program, the program would fail to reduce GHG emissions. Another concern is whether the inclusion of offsets would send the appropriate price signal to encourage the development of long-term mitigation technologies. Policymakers may consider a balance between price signal and program costs.

If eligible in a U.S. program, international offsets are expected to dominate in early decades because they would likely offer the lowest-cost options. Domestic sectors, such as agriculture and forestry, might benefit if international offsets are excluded. Some object to the use of international offsets due to concerns of fairness: the low-cost options would be unavailable to developing nations if and when they establish GHG emission targets. However, some offset projects may promote sustainable development. On the other hand, international offsets may serve as a disincentive for developing nations to enact laws or regulations controlling GHG emissions because many projects would no longer qualify as offsets.

INTRODUCTION

A variety of efforts to address climate change are currently underway or being developed on the international, national, and sub-national levels (e.g., individual state actions or regional partnerships).[1] These efforts cover a wide spectrum, from climate change research to mandatory greenhouse gas (GHG) emissions reduction programs.[2] In Congress, Members introduced a number of proposals that would establish a national GHG emissions reduction regime.

GHG emissions reduction programs, both ongoing and proposed, vary considerably. The primary variables are scope and stringency: which emission sources are covered by the program and how much emission reduction is required.[3] These factors largely determine the impacts of an emissions reduction program, but other design details can have substantive effects.

One such design element is the treatment of offsets. An offset is a measurable reduction, avoidance, or sequestration of GHG emissions from a source *not covered* by an emission reduction program. If a cap-and-trade program includes offsets, regulated entities have the opportunity to purchase them to help meet compliance obligations.[4]

Offsets have generated debate and controversy in climate change policy. If Congress establishes a federal program to manage or reduce GHG emissions, whether and how to address offsets would likely be an important issue. Because most current and proposed programs allow offsets (see tables at the end of the report), offset projects will probably play some part in an emissions reduction program.

The first section of this report provides an overview of offsets by discussing different types of offset projects and describing how the offsets would likely be used in an emission reduction program. The next section discusses the supply of offsets that might be available in an emission trading program. The subsequent sections examine the potential offset benefits and the potential concerns associated with offsets. The final section offers considerations for Congress. In addition, the report includes a table comparing the role of offsets in selected emission reduction programs: proposals in the 111th and 110th Congresses, U.S. state initiatives, and international programs.

OFFSETS: AN OVERVIEW

Offsets are sometimes described as project-based because they typically involve specific projects or activities whose primary objective is to reduce, avoid, or sequester emissions. Because offset projects can involve different GHGs, they are quantified and described with a standard form of measure: either metric tons of carbon-equivalents (mtC-e) or metric tons of CO_2-equivalents (mtCO_2-e).[5]

To be credible as offsets, the emissions reduced, avoided, or sequestered must be *additional* to business-as-usual (i.e., what would have happened anyway). This concept is often called "additionality." If Congress establishes a GHG emission cap-and-trade program, only sources not covered by the cap could generate offsets.[6] Emission reductions from regulated sources (e.g., coal-fired power plants) would either be required or spurred by the emissions cap.[7] In contrast, if agricultural operations were not covered under an emissions cap, a project that collects methane emissions from a manure digester would likely be an *additional* GHG emission reduction.

If offsets are allowed as a compliance option in an emissions trading program, eligible offset projects could generate "emission credits," which could be sold and then used by a regulated entity to comply with its reduction requirement.[8] This approach is part of the European Union's (EU) Emission Trading Scheme (ETS), which EU members use to help meet their Kyoto Protocol commitments.[9] Under the EU ETS, regulated entities can purchase emission credits that are created from approved offset projects.[10] Regulated entities can then apply the credits towards their individual emission allowance obligations.[11] For example, a regulated entity may consider purchasing offsets if the offsets are less expensive than making direct, onsite emission reductions. Assuming the offset is legitimate—i.e., a ton of carbon reduced, avoided, or sequestered through an offset project equates to a ton reduced at a regulated source—the objective to reduce GHG emissions is met. From a global climate change perspective, it does not matter where or from what source the reduction occurs: the effect on the atmospheric concentration of GHGs would be the same.

Offsets increase emission reduction opportunities. When offsets are not allowed, incentives to reduce emissions or sequester carbon are limited to the covered sources, and there is little motivation to improve mitigation technologies for non-covered sources. Including offsets in a cap-and-trade program would expand these incentives.

Voluntary Offsets

Although this report focuses on the use of offsets in a mandatory GHG emissions reduction program, offsets are generating interest and debate in other contexts. In the United States and around the world, a growing number of businesses, interest groups, and individuals are purchasing offsets and claiming that all or part of their GHG-emitting activities (e.g., travel or specific events) are "carbon neutral." The motivation for these purchases can vary. Some businesses may be seeking to enhance their public image; others may hope to take credit for the offsets in a future GHG reduction program. The exchanges represent a voluntary market for offsets because there is no requirement for the parties to curtail their emissions.

The quality of the offsets in the voluntary market varies considerably, largely because there are no commonly accepted standards. Some offset sellers offer offsets that comply with the more explicit standards of the Kyoto Protocol's Clean Development Mechanism. Other sellers offer offsets that meet the seller's self-established guidelines, which may be considered proprietary information, and thus not publicly available.

Due to the lack of common standards, some observers have referred to the market as the "wild west." This should not suggest that all offsets are low quality, but that the consumer must adopt a buyer-beware mentality when purchasing offsets. For more information, see CRS Report RL34241, *Voluntary Carbon Offsets: Overview and Assessment*, by Jonathan L. Ramseur.

Offset Types and Examples

Offsets could potentially be generated from an activity that emits GHGs or that would remove or sequester GHGs from the atmosphere. This section discusses offsets in four categories. Each category is discussed below with project examples for each group.

Some of the categories and examples listed below may be limited by location. If a U.S. law or regulation (other than an emissions cap) governs a specific emission source (e.g., methane from coal mines), that source's emission reductions would not qualify as domestic offsets, unless the reductions made went further than the regulations required.[12] For example, if the source is required by law or regulation to reduce methane emissions by 50%, reductions up to this threshold would *not* qualify as offsets, but reductions in excess of 50% might qualify as offsets. As more nations establish mandatory caps or require specific technological controls or practices at emission sources, the universe of potential offsets would shrink.

Biological Sequestration

Trees, plants, and soils sequester carbon, removing it from the earth's atmosphere. Biological sequestration projects generally involve activities that either increase existing sequestration; or maintain the existing sequestration on land that might otherwise be disturbed and release some or all of the sequestered carbon. This offset category includes sequestration that results from agriculture and forestry activities, and is sometimes referred to as land use, land use change and forestry (LULUCF) projects. Examples of these projects include

- planting trees on previously non-forested land (i.e., afforestation);
- planting trees on formerly forested land (i.e., reforestation);
- limiting deforestation by purchasing forested property and preserving the forests with legal and enforcement mechanisms;
- setting aside croplands from agricultural production to rebuild carbon in the soil and vegetation; and
- promoting practices that reduce soil disruption: e.g., conservation tillage and erosion control.[13]

Compared to the other offset categories discussed here, biological sequestration projects, particularly forestry projects, offer the most potential in terms of volume. However, this category is arguably the most controversial because several integrity issues are typically (or perceived to be) associated with biological sequestration projects.[14] These issues are discussed in more detail in later sections of this report.

Renewable Energy Projects

Historically, renewable energy—e.g., wind, solar, biomass—has been a more expensive source of energy than fossil fuels.[15] A renewable energy offset project could provide the financial support to make renewable energy sources more economically competitive with fossil fuels. Renewable energy sources generate fewer GHG emissions than fossil fuels, particularly coal. Wind and solar energy produce zero direct emissions. Use of renewable sources would avoid emissions that would have been generated by fossil fuel combustion. These avoided emissions could be sold as offsets. Potential renewable energy offset projects may include[16]

- constructing wind farms to generate electricity;
- adding solar panels;
- retrofitting boilers to accommodate biomass fuels; and
- installing methane digesters at livestock operations.[17]

Domestic renewable energy projects are not likely to qualify as offsets in a national emissions reduction program. In a carbon-constrained context, project developers would be hard-pressed to demonstrate that a renewable energy project would not have happened anyway. In an "economy-wide" cap-and-trade emissions program, energy sector emissions would likely be capped. The cap would make fossil fuels more expensive and renewable energy sources more attractive. However, renewable energy projects may still create credible offsets in nations without GHG emission controls on their energy sectors.

Energy Efficiency

A more energy efficient product or system requires less energy to generate the same output. Improvements in energy efficiency generally require a financial investment in a new product or system. These capital investments likely pay off in the long run, but the payback period may be too long or capital financing may be constrained, particularly for small businesses or in developing nations. Examples of possible energy efficiency offset projects include

- upgrading to more efficient machines or appliances;
- supporting construction of more energy efficient buildings;
- replacing incandescent light bulbs with fluorescent bulbs.

Similar to renewable energy offsets, domestic energy efficiency offset projects would likely face substantial hurdles in proving their additionality in a carbon-constrained regime. As the price of carbon increases and raises energy prices—both outcomes expected with an emissions cap—the incentive to reduce energy use through energy efficiency improvements will increase.

Offset ownership is another potential challenge regarding some energy efficiency offsets. Energy efficiency improvements may occur at a different location than the actual reduction in emissions. For example, a business that runs its operations with purchased electricity will use less electricity if energy efficiency improvements are made, but the actual emission reductions will be seen at a power plant. Thus, the reductions may be counted twice: first as an energy efficiency offset and second as a direct reduction at the power plant. One way to address this potential dilemma is to restrict energy efficiency projects to only those that reduce or avoid *on-site* combustion of fossil fuels. This approach is used in the few congressional proposals that specifically allow energy efficiency offsets.

As with renewable energy projects, there could be energy efficiency projects in nations that do not limit GHG emissions.

Non-CO_2 Emissions Reduction

Multiple sources emit non-CO_2 greenhouse gases. These emissions are often not controlled through law or regulation. These sources—primarily, agricultural, industrial, and waste management facilities—emit GHGs as by-products during normal operations. In many cases, the individual sources emit relatively small volumes of gases. However, there are a large number of individual sources worldwide, and many of the gases emitted have greater global warming potential (GWP) than carbon dioxide.[18] Offset projects in this category would generally provide funding for emission control technology to reduce these GHG emissions. Examples of emission reduction opportunities include the following:

- methane (CH_4) emissions from landfills, livestock operations, or coal mines (GWP = 25)
- nitrous oxide (N_2O) emissions from agricultural operations or specific industrial processes (GWP = 298)
- hydrofluorocarbon (HFC) emissions from specific industrial processes, such as HFC-23 emissions from production of a refrigerant gas (GWP of = 14,800)
- sulfur hexafluoride (SF_6) from specific industrial activities, such as manufacturing of semiconductors (GWP = 22,800)

This offset category is broad, as it involves many different industrial activities. As such, some offset types in this category are generally considered high quality, and others that have generated controversy. For example, methane reduction from landfills or coal mines has a reputation as a high quality offset. These projects are relatively easy to measure and verify, and in many cases would likely not occur if not for the financing provided by an offset market. Therefore, the challenge of proving additionality is easier to overcome.

Offsets involving abatement of HFC-23 emissions from production of a common refrigerant[19] have spurred controversy. Of the offset types certified through the Kyoto Protocol's Clean Development Mechanism (CDM), HFC-23 offsets represent the greatest percentage: 50% of the certified emission reductions (CERs) have come from HFC-23 abatement projects.[20] Controversy has arisen because the production facilities can potentially earn more money from the offsets (destroying HFC-23 emissions) than from selling the primary material.[21] This creates a perverse incentive to produce artificially high amounts of product to generate a more lucrative by-product.

Policy Alternatives to Offsets

Policymakers have alternative methods of addressing the emission sources and sinks that are often considered to be candidates for offsets.

Emissions Cap. Congress could expand the scope of the emissions cap to include emission sources that were previously excluded. The rationale for initially excluding these sources is that they are large in number, and they individually generate a relatively small quantity of emissions. Therefore, an offset program is arguably a more cost-effective means of achieving reductions from these sources. However, including certain sources, while excluding others, may raise issues of fairness. For example, some may question why specific sources are capped, while other sources can generate financial gain through the offset market. This discussion is beyond the scope of this report.

Emissions Standards. Instead of allowing offsets from non-capped sources, Congress could establish sector-specific emission performance standards or technological requirements. This approach is sometimes described as "commandand-control." Such a policy could be applied to both emission sources and sequestration activities. If Congress sets a baseline requirement, reductions or sequestration beyond the minimum requirement could qualify as offsets.

Set-Aside Allowances. One possible design element of an emission control program is for policymakers to allot a specific percentage of emission allowances ("set-asides") from the overall emissions cap to non-regulated entities (i.e., parties not subject to the emissions cap). In a carbon-constrained context, the set-aside allowances are essentially currency because they could be sold to regulated facilities to help meet compliance obligations. Set-asides can be allocated to parties to promote various objectives, including support for activities that reduce, avoid, or sequester emissions. The allowances may also be distributed for other purposes, such as transition assistance to specific economic sectors or financial support to low-income households. These groups may pay proportionately higher costs in an emission reduction regime.

Although both set-aside allowances and offsets would address emissions in sectors not subject to the cap, their impacts on regulated sources would differ substantially. Set-aside allowances are within the emissions cap. Offsets represent compliance options from sources outside of the cap. Neither offsets nor set-asides would alter the GHG reduction goal of the program: the cap would remain the same.

However, offsets would increase the emission reduction opportunities available to regulated sources; set-aside allowances would not.

Potential Benefits of Offsets

The inclusion of offsets in a cap-and-trade program could potentially provide multiple benefits. Perhaps the primary benefit would be improved cost-effectiveness. The ability to generate offsets, which could be sold as emission credits, would provide an incentive for non-regulated sources to reduce, avoid, or sequester emissions. The inclusion of offsets could expand emission mitigation opportunities, likely reducing compliance costs for regulated entities. Many offset projects have the potential to offer environmental benefits, as well. Developing countries, in particular, may gain if the United States includes international offsets in a GHG emission program. In addition, the offset market may create new economic opportunities and spur innovation as parties seek new methods of generating offsets. These issues are discussed below in greater detail.

Cost-Effectiveness

A central argument in support of offsets is that their use makes an emissions reduction program more cost-effective. A wide range of activities could be undertaken that would generate offsets. Many of these individual activities would likely generate a relatively small quantity of offsets (in terms of tons), but in the aggregate, their climate change mitigation potential is substantial. Arguably, direct regulation of these sources—either through a cap-and-trade program or funding for emission control technology to reduce these GHG emissions. Examples of emission reduction opportunities include the following:

- methane (CH_4) emissions from landfills, livestock operations, or coal mines (GWP = 25)
- nitrous oxide (N_2O) emissions from agricultural operations or specific industrial processes (GWP = 298)
- hydrofluorocarbon (HFC) emissions from specific industrial processes, such as HFC-23 emissions from production of a refrigerant gas (GWP of = 14,800)

- sulfur hexafluoride (SF_6) from specific industrial activities, such as manufacturing of semiconductors (GWP = 22,800)

This offset category is broad, as it involves many different industrial activities. As such, some offset types in this category are generally considered high quality, and others that have generated controversy. For example, methane reduction from landfills or coal mines has a reputation as a high quality offset. These projects are relatively easy to measure and verify, and in many cases would likely not occur if not for the financing provided by an offset market. Therefore, the challenge of proving additionality is easier to overcome.

Offsets involving abatement of HFC-23 emissions from production of a common refrigerant[19] have spurred controversy. Of the offset types certified through the Kyoto Protocol's Clean Development Mechanism (CDM), HFC-23 offsets represent the greatest percentage: 50% of the certified emission reductions (CERs) have come from HFC-23 abatement projects.[20] Controversy has arisen because the production facilities can potentially earn more money from the offsets (destroying HFC-23 emissions) than from selling the primary material.[21] This creates a perverse incentive to produce artificially high amounts of product to generate a more lucrative by-product.

The study also found that international offsets would play a large role, especially in the beginning decades of the program, because there are generally more low-cost offset opportunities in other nations. In later years (as the carbon price rises), domestic offset types, particularly forestry-related offsets, play a larger role.

Potential Co-Benefits

Offset projects may produce benefits that are not directly related to climate change. For example, many of the offset projects that promote carbon sequestration in soil (e.g., conservation tillage) improve soil structure and help prevent erosion.[25] Erosion control may reduce water pollution from nonpoint sources,[26] a leading source of water pollution in U.S. waterbodies.[27]

Depending on a project's specific design and how it is implemented, other agriculture and forestry offset projects could potentially yield positive environmental benefits. However, there is some concern that certain projects may produce undesirable impacts, such as depleted soil quality, increased water use, or loss of biodiversity.[28] Many agriculture and forestry offset projects would likely involve land use changes, such as converting farmlands to forests or biofuel production.[29] Determining whether the change imparts net benefits may be a complex evaluation, depending upon, among other things, the current and proposed species of plants and/or trees. Policymakers would likely encounter projects that offer trade-offs: for example, they offset GHG emissions, while imposing an unwanted outcome, such as increased water use, reducing availability downstream.[30] EPA found that the more aggressive offset opportunities—afforestation and biofuels production—are more likely to present the most distinct trade-offs.[31]

Potential Benefits to Developing Nations

Most observers would agree that developing nations are unlikely to limit and reduce GHG emissions on a schedule on par with developed nations. With less-regulated emission sources, the universe of eligible offset opportunities would be much larger in developing nations. Offset types, such as renewable energy and/or energy efficiency projects, which could face substantial hurdles to qualify as offsets in the United States, would be eligible offsets from developing nations. These types of projects would likely provide environmental benefits beyond GHG emission reduction— improvements in local air quality—by displacing or avoiding combustion of fossil fuels.

Offset projects in developing nations have the potential to promote sustainable development, such as creation of an energy infrastructure that is less carbon-intensive and more energy efficient. In fact, this was one of the objectives in establishing the Clean Development Mechanism (CDM). Whether this objective is being met is a subject of debate. However, recent projections suggest that offset activities that promote sustainable development will account for a larger percentage of emissions credits in the coming years.

As a comparison between Figure 2 and Figure 3 indicates, the proportion of renewable energy and energy efficiency projects in the CDM is expected to more than double by 2012. This projected shift would likely improve support for sustainable development objectives. However, offset projects—primarily, HFC and N2O reduction from industrial activities—that provide few sustainable development benefits are still expected to account for a considerable proportion of emission credits issued.

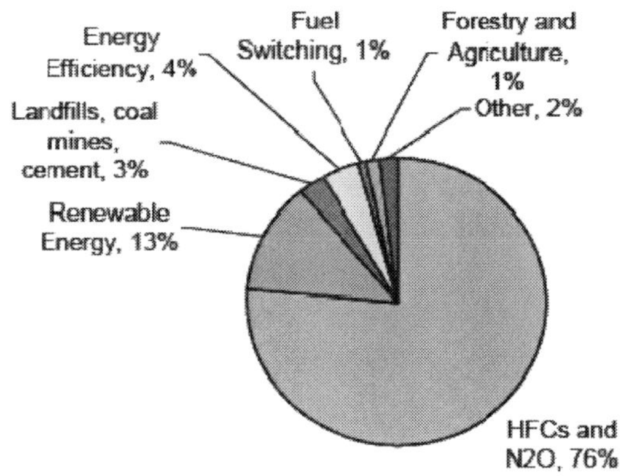

Figure 2. CERs Issued to Data by Project Type (as of May 1, 2009).

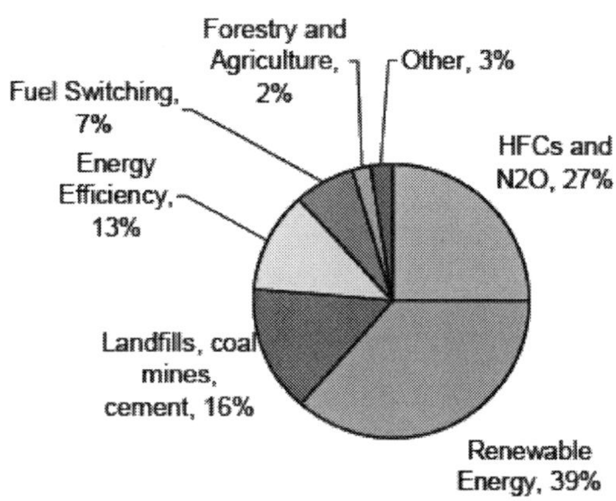

Source for Figure 2 and Figure 3: Prepared by CRS with data from United Nations Environment Programme, Capacity Development for the Clean Development Mechanism ("CDM Pipeline"), at http://cd4cdm.org/index.htm.

Figure 3. 2012 Projections for CERs by Project Type (as of May 1, 2009).

Other Potential Domestic Benefits

A federal cap-and-trade emission program that allows offsets as a compliance option may provide economic benefits to particular sectors of the U.S. economy. However, there may be trade-offs, depending on which types of offsets are eligible and whether or not international offsets are allowed. If international offset projects are included in the program, some U.S. business sectors may benefit from the transfer of technology and/or services to support projects in other nations. If international offsets, generally the lowest-cost options, are excluded, the offset projects from the domestic agriculture and forestry sectors would likely gain a greater share of the offsets market, thus generating business opportunities in these sectors.[32]

Another potential benefit that is often highlighted is the ability of an offset market to encourage innovation. As the carbon price provides an incentive for regulated entities to find onsite emission reductions (e.g., through efficiency improvements or development of new technologies), the offset market may spur parties to find new ways to reduce, avoid, or sequester emissions from non-regulated sources. However, there is some concern that the drive to find creative offset methods may encourage offset projects that yield unknown, unintended, and possibly harmful, environmental effects. A frequently cited example in this regard is ocean fertilization, which seeks to stimulate phytoplankton growth (and ultimately improve CO_2 sequestration) by releasing iron into certain parts of the surface ocean.[33]

POTENTIAL CONCERNS

Although offsets have the potential to provide benefits under an emissions trading program, several issues associated with offsets have generated concern and some controversy. Perhaps the primary concern regarding offsets is their integrity. To be credible, an offset should equate to an emission reduction from a direct emission source, such as a smokestack or exhaust pipe. This issue is critical, if offsets are to be used in an emissions trading program. However, implementing this objective would likely present challenges. This and other concerns are discussed below.

Supplementarity

Supplementarity refers to the idea that the role of offsets in an emission reduction program should be secondary to reduction efforts at regulated emission sources. The term comes from the text of the Kyoto Protocol, which states that emissions credits (or offsets) must be "*supplemental* to domestic actions for the purpose of meeting quantified emission limitations and reduction commitments.... " (Article 17, emphasis added).

Integrity Concerns

If offsets are to be included in an emissions trading program, offset integrity—i.e., whether or not the offsets represent real emission reductions—is critical.

Several issues need to be addressed when evaluating offsets. Some of these issues may present implementation challenges, which if not overcome, could damage the integrity of the offset. These issues are discussed below.

Additionality

Additionality means that the offset project represents an activity that is beyond what would have occurred under a business-as-usual scenario. In other words, would the emission reductions or sequestration have happened anyway?

Additionality is generally considered to be the most significant factor that determines the integrity of the offset. In the context of an emissions control program, a test of additionality would examine whether the offset project would have gone forward in the absence of the program. An additionality determination would likely consider the following questions:

- Does the activity represent a common practice or conforms to an industry standard?
- Is the offset project required under other federal, state, or local laws?
- Would the project generate financial gain (e.g., be profitable) due to revenues from outside the offset market?[34]

Offset credits allow regulated entities to generate GHG emissions above individual compliance obligations. If project developers are able to generate emission credits for projects that would have occurred regardless (i.e., in the absence of the trading program), the influx of these credits into the program would undermine the emissions cap and the value of other, legitimate offset projects.

Additionality is at the crux of an offset's integrity, but applying the additionality criterion may present practical challenges. For instance, it may be impossible to accurately determine "what would have happened anyway" for some projects.

Assessing a project's additionality may involve some degree of subjectivity, which may lead to inconsistent additionality determinations.

"Stackability" and Additionality

The concept of "stackability" is typically discussed in the context of biologicial sequestration offset projects. In addition to decreasing atmospheric concentrations of GHGs, biological sequestration projects may provide nonclimate-related ecosystem services, such as improved water quality and wildlife habitat enhancement. Proponents of "stackability" argue that offset project developers should be able to market these services separately and earn distinct financial benefits for each ecosystem service (assuming a funding source exists that would support each service). Under this reasoning, the financial rewards for different ecoystem services would be "stackable," and the receipt of funding from one source (e.g., buyers in the offset marketplace) would not preclude the receipt of funding from another source (e.g., government grant program). For example, some argue that a soil sequestration project should be able to generate offset credits for its GHG sequestration services and a Conservation Reserve Program (CRP) payment for the project's ability to provide erosion control.

Allowing an offset project to accrue stackable benefits may raise concerns of additionality. If non-climate-related incentive programs or ecosystem service markets could provide financial support for a particular offset project, observers may question whether the project would have happened anyway. In some situations, an additionality assessment may be relatively straightforward. If the non-climate-related incentives stimulate activities that would mitigate GHG emissions (as a secondary effect) without the support of the offset market, the activity would not likely qualify as additional in terms of carbon offsets. On the other hand, some offset projects may not be economically viable without multiple sources of funding—combining a payment from the offset market with grants from nonclimate-related government programs. Thus, in some situations a determination of additionality may entail a degree of subjectivity.

Measurement

Reliable GHG emissions data are a keystone component of any climate change program. If Congress allows offsets as a compliance option, offset data (emissions reduced, avoided, or sequestered) should arguably be as reliable as data from regulated sources. From a practical standpoint, however, achieving this objective may be difficult.

It is generally much simpler to measure and quantify an emission reduction from a direct source than from an offset project. Indeed, the more difficult measurement may be the main reason such reductions are not required by a control program. Regulated sources determine their compliance by comparing actual GHG emissions data against their allowed emissions.[35] In contrast, project developers determine offset emission data by comparing the expected reduced, avoided, or sequestered GHG emissions against a projected, business-as-usual scenario (sometimes referred to as a counter-factual scenario).

To accomplish this task, offset project managers must establish an emissions baseline: an estimate of the "business-as-usual" scenario or the emissions that would have occurred without the project. If project managers inaccurately estimate the baseline, the offsets sold may not match the actual reductions achieved. For example, an overestimated baseline would generate an artificially high amount of offsets. Baseline estimation may present technical challenges. In addition, project developers have a financial incentive to err on the high side of the baseline determination because the higher the projected baseline, the more offsets generated. Requiring third-party verification (as some proposals do) would potentially address this specific concern.

Biological sequestration offset projects may present particular challenges in terms of measurement. The carbon cycle in trees and soils is only partially understood.[36] Variations exist across tree species, ages, soil conditions, geographic locations, and management practices.[37] Estimates of carbon uptake and storage are frequently considered imprecise or unreliable. Further, changes in vegetation cover may have non-emission effects on climate, such as how much of the Sun's energy is reflected or absorbed by the Earth. A recent study in the *Proceedings of the National Academy of Sciences* stated, "Latitude-specific deforestation experiments indicate that afforestation projects in the tropics would be clearly beneficial in mitigating global-scale warming, but would be counterproductive if implemented at high latitudes and would offer only marginal benefits in temperate regions."[38]

Double-Counting

To be credible, when an offset is sold, it should be retired and not sold again or counted in other contexts. However, opportunities for double-counting exist. For example, a regulated entity may purchase offsets generated through the development of a wind farm in a nation that has not established GHG emissions targets. The U.S. buyer would count the offsets, which may have been purchased to negate increased, onsite emissions at the regulated source. In addition, the nation, in which the wind farm is located, would likely see an emissions reduction due to the wind farm. If this decrease is reflected in the nation's GHG emissions inventory, the offset project (wind farm) might replace other reduction activities that the nation might have taken to meet its target.

Some may argue that double-counting is less of a problem if the offset project occurs in a nation with only a voluntary target (as opposed to a nation subject the Kyoto Protocol). However, the impact would be the same if the nation eventually establishes a mandatory target and takes credit for the earlier reductions associated with the offset project. By taking credit for an earlier reduction, the nation might need to make fewer reductions to be in compliance with the new mandatory program.

A tracking system could help avoid such double-counting.[39] Most would agree that a domestic tracking system would be simpler to establish and monitor than a system that follows international offset trading. The latter would require, at a minimum, cooperation with the nations hosting the offset projects.

Permanence

With some offset projects there may be a concern that the emission offsets will be subsequently negated by human activity (e.g., change in land use) or a natural occurrence (e.g., forest fire, disease, or pestilence). This issue is most pertinent to biological sequestration projects, specifically forestry activities. Although many observers expected forestry offsets to play a large role in the CDM, this has not been observed in practice. This result is partially due to concerns of offset permanence in developing nations.[40]

Offset buyers need some assurance that the land set aside for forests (and carbon sequestration) will not be used for a conflicting purpose (e.g., logging or urban development) in the future. Although natural events (fires or pests) are hard to control, human activity can be constrained through legal documents, such as land easements. In addition, an offset could come with a guarantee that it would be replaced if the initial reduction is temporary. Permanence may be more difficult to monitor at international projects.

Leakage

In the context of climate change policy, GHG emissions leakage generally refers to a situation in which an emissions decrease from a regulated (i.e., capped) source leads to an emissions increase from an unregulated source. EPA states that leakage "occurs when economic activity is shifted as a result of the emission control regulation and, as a result, emission abatement achieved in one location that is subject to emission control regulation is [diminished] by increased emissions in unregulated locations."[41]

Leakage scenarios may involve emission sources from the same economic sector, but located in different countries.

Many voice concern that if the United States were to cap emissions from specific domestic industries (e.g., cement, paper), these industries would relocate to nations without emission caps and increase activity (and thus emissions) to compensate for the decreased productivity in the United States. Thus, global net emissions would not decrease, and affected domestic industries would likely see employment losses.

In the context of offsets, leakage may occur in an analogous fashion. The opportunity for leakage exists when an offset project decreases the supply of a good in one location, leading to greater production of the good somewhere else.

Compared to other offset types, forestry projects, particularly those that sequester carbon by curbing logging, likely present the greatest risk of leakage.[42] For example, an offset project that restricts timber harvesting at a specific site may boost logging at an alternative location, thus reducing the effectiveness of the offset project.[43] Preventing or accounting for leakage from these projects poses a challenge.

Delay of Technology Development

As discussed above, the inclusion of offsets would likely lower the overall cost of compliance. Although many consider this a desired outcome, some contend that the price of carbon needs to reach levels high enough to promote the long-term technological changes needed to mitigate climate change.

> Offsets also can delay key industries' investments in transformative technologies that are necessary to meet the declining cap. For instance, unlimited availability of offsets could lead utilities to build high-emitting coal plants instead of investing in efficiency, renewables, or plants equipped with carbon capture and storage.[44]

Transaction Costs

Transaction costs generally refer to the costs associated with an exchange of goods or services. In an offset market, transaction costs may encompass the following:

- searching for offset opportunities;
- studying and/or measuring offset projects;
- negotiating contracts;
- monitoring and verifying reduced, avoided, or sequestered emissions;
- seeking regulatory approval;
- obtaining insurance to cover risk of reversal (i.e., non-permanence).[45]

Depending on the price of carbon in the offset market, transaction costs may represent a substantial percentage of the value of the offset. Several studies have examined offset projects in an effort to estimate transaction costs. Generally, the studies' results include a transaction cost range that varies by offset type and project size. For example, a study by the Lawrence Berkeley National Laboratory (LBL) found a transaction cost range of \$0.03/mtCO_2-e to \$4.05/mtCO_2-e.[46] Overall, the various studies found that smaller offset projects (measured by tons of CO_2-e) may be at a disadvantage because they would likely face proportionately higher transaction costs: the LBL study found that the mean transaction cost for small projects was \$2.00/mtCO_2-e, but only \$0.35/mtCO_2-e for the largest projects.

The transaction costs may hinder innovation by serving as an obstacle to small, but promising offset projects. However, transaction costs are inherent in an emissions program that requires project developments to meet certain provisions—additionality, measurement, verification, monitoring—to maintain the integrity of the offset allowed as compliance alternatives.

Concerns in Developing Nations

Some argue that offset use, particularly unlimited access to international offset opportunities, raises questions of fairness. Most of the world's GHG emissions (especially on a per capita basis) are generated in the developed nations, while most of the lower-cost offset opportunities are in developing nations. Many observers expect the developing nations to establish mandatory GHG reduction programs several years (if not decades) after developed nations' emission programs are underway. The developed nations are likely to initiate the lower-cost projects and retire the offsets, thus removing the "low-hanging fruit." If and when the developing nations subsequently establish GHG emission caps, the lower-cost compliance alternatives would not be available to them.[47] Some have described this as a form of environmental colonialism.[48]

Another concern is that international offsets may serve as a disincentive for developing nations to enact laws or regulations limiting GHG emissions. For instance, if a developing nation established emission caps or crafted regulations for particular emissions sources, reductions from these sources would no longer qualify as offsets. Developing nations may be hesitant to forego the funding provided by offset projects.

CONSIDERATIONS FOR CONGRESS

From a climate change perspective, the location of an emission activity does not matter: a ton of CO_2 (or its equivalent in another GHG) reduced in the United States and a ton sequestered in another nation would have the same result on the atmospheric concentration of GHGs. Moreover, unlike many air pollutants—e.g., acid rain precursors sulfur dioxide and nitrogen oxide, particulate matter, and mercury—a localized increase or decrease of CO_2 emissions does not directly impart corresponding local or regional consequences. This attribute of CO_2 emissions, the primary GHG, allows for offset opportunities.[49]

If allowed as part of an emissions reduction program, offsets have the potential to provide various benefits. The ability to generate offsets may

- provide an incentive for non-regulated sources to reduce, avoid, or sequester emissions (where these actions would not have occurred if not for the offset program);
- expand emission mitigation opportunities, thus reducing compliance costs for regulated entities;
- offer environmental co-benefits for certain projects;
- support sustainable development in developing nations; and
- create new economic opportunities and spur parties to seek new methods of generating offsets.

The main concern with offset projects is whether or not they produce their stated emission reductions. To be credible, an offset ton should equate to a ton reduced from a direct emission source, such as a smokestack or exhaust pipe. If offset projects generate emission credits for activities that would have occurred anyway (i.e., in the absence of the emission trading program), these credits would not satisfy the principle of additionality. For many offset projects, determining additionality will likely pose a challenge. Other offset implementation issues— baseline estimation, permanence, accounting, monitoring—may present difficulties as well. If illegitimate offset credits flow into the trading program, the cap would effectively expand and credible emissions reductions would be undermined. The program would fail to meets its ultimate objective: overall GHG emissions reductions.

Offset projects vary by the quantity of emission credits they could generate and the implementation complexity they present. For instance, domestic landfill methane projects are comparatively simple to measure and verify, but offer a relatively small quantity of offsets. In contrast, biological sequestration activities, particularly forestry projects, offer the most offset-generating potential, but many of these projects pose multiple implementation challenges. This may create a tension for policymakers, who might want to include the offset projects that provide the most emission reduction opportunities, while minimizing the use of offset projects that pose more implementation complications.[50] Addressing these challenges may require independent auditing and/or an appreciable level of oversight and administrative support from government agencies. A report from the National Commission on Energy Policy stated, "Proposals that expect to achieve significant (> 10 percent) compliance through offsets in the near term will be obligated to create a substantial enforcement bureaucracy or risk an influx of illegitimate credits."[51]

If concerns of legitimacy can be resolved, the next question for policymakers may be whether the potential benefits provided by offsets would outweigh any potential harm. One debate may involve whether including offsets would send the appropriate price signal to encourage the development and deployment of new technologies, such as carbon capture and storage. Policymakers may consider striking a balance between sending a strong price signal and reducing the costs of the emissions reduction program.

Table 1. Comparison of Offset Treatment in GHG Emission Control Proposals in the 111th Congress

GHG Reduction Program	Offset Types Allowed or Prohibited	Offset Quantity Limitation	Acceptable Locations of Offset Projects	System of Verifying Integrity of Offsets
H.R. 1666 (Doggett)	Offsets are not included in the bill	Offsets are not included in the bill	Offsets are not included in the bill	Offsets are not included in the bill
H.R. 1862 (Van Hollen)	Offsets are not included in the bill	Offsets are not included in the bill	Offsets are not included in the bill	Offsets are not included in the bill
H.R. 2454 (Waxman/Markey)	Eligible domestic offset types to be determined through EPA rulemaking process International offsets may be generated (1) on project-by-project basis (subject to certain conditions), but allowable projected not identified; (2) on a sectoral basis; and (3) from avoided deforestation efforts	In 2012, approximately 30% of an entity's allowance obligation can be satisfied with offsets; this percentage increases to 67% by 2050; if all entities maximized their use of offsets, the aggregate annual number of submitted offsets would total 2 billion tons Half of an entity's offsets can come from domestic sources and half from international sources (e.g., 15% domestic and 15% international in 2012); EPA can increase the allowable percentage for international offsets, if the agency determines use of domestic offsets will not be maximized in a particular year In 2018, international offsets are discounted: 1.25 offsets equals 1 emission allowance	Domestic and international	Directs the EPA to develop regulations to implement offset program; EPA is to develop methodologies—protocols for additionality, measurement, etc.—for each offset type that the agency deems eligible; EPA must establish policies for addressing reversals; Offset project developers to submit a petition to EPA and receive approval of project; offset projects must then be reviewed by an accredited third-party, who submits report to EPA for approval; EPA is to conduct periodic audits of offset projects

Note: Other GHG emission control programs introduced in the 111th Congress would use a price-control approach (e.g., carbon tax) to reduce GHG emissions. Tax credits (somewhat analogous to offsets in a cap-and-trade program) for emission reduction/sequestration projects from non-covered sources are not included in these proposals. For more details regarding these proposals and the proposals listed above, see CRS Report R40556, *Market-Based Greenhouse Gas Control: Selected Proposals in the 111th Congress*, by Jonathan L. Ramseur, Larry Parker, and Brent D. Yacobucci.

Table 2. Comparison of Offset Treatment in GHG Emission Control Proposals from the 110th Congress

GHG Reduction Program	Offset Types Allowed or Prohibited	Offset Quantity Limitation	Acceptable Locations of Offset Projects	System of Verifying Integrity of Offsets
S. 280 (Lieberman)	Specifically allows for agricultural and conservation practices; reforestation; forest preservation; directs EPA to determine other offset types	Up to 30% of allowances can come from domestic or international offsets; if offsets account for 15% of allowances, at least 1.5% must come from agricultural sequestration	Domestic or international projects accepted	Directs the EPA Administrator, in coordination with the Secretaries of Commerce, Energy, and Agriculture, to set standards
S. 309 (Sanders)	Offsets are not included in the bill	Offsets are not included in the bill	Offsets are not included in the bill	Offsets are not included in the bill Directs EPA to implement emissions reduction program; directs Secretary of Agriculture, in coordination with EPA, to develop standards for biological sequestration
S. 317 (Feinstein) (Covers utilities)	No specific prohibitions; specifically allows biological sequestration, including agricultural and forestry activities, and emission reductions from various industrial operations;[a] EPA may allow further types at its discretion	A regulated source can use domestic offsets to cover 100% of its allowances; up to 25% of allowances (50% for new sources) can come from eligible international offsets; this limit increases to 50% if allowance prices reach a level that would cause "significant harm" to the economy (as determined by EPA and Secretary of Treasury)	Domestic and international projects allowed	Directs Secretary of Agriculture, in coordination with EPA, to develop standards for biological sequestration offsets; directs EPA to craft standards for other project types

GHG Reduction Program	Offset Types Allowed or Prohibited	Offset Quantity Limitation	Acceptable Locations of Offset Projects	System of Verifying Integrity of Offsets
S. 485 (Kerry) S. 1168 (Alexander) (CO_2 reduction from utilities)	Offsets are not included in the bill Five offset types allowed: 1) landfill methane reduction 2) sulfur hexafluoride (SF6) reductions from industrial activities 3) afforestation projects 4) energy efficiency projects yielding reductions or avoidance of CO_2 from natural gas, oil or propane combustion 5) avoided methane from manure management practices	Offsets are not included in the bill No quantity limitations	Offsets are not included in the bill Any U.S. state that has signed memorandum of understanding (MOU) with EPA	Offsets are not included in the bill Directs Secretary of Agriculture, in coordination with EPA, to develop standards for biological sequestration Legislation includes specific standards[b] for the five offset types allowed; directs EPA to develop standards for other potential project types, including agricultural offsets
S. 1177 (Carper) (CO_2 reduction from utilities)	Identifies 11 eligible types, including agricultural and forestry management practices; authorizes EPA to develop standards for additional types	No limits; directs EPA to develop regulations regarding use of offsets	Directs EPA to develop standards for domestic and international locations	Directs EPA to develop regulations and coordinate with Department of Agriculture regarding biological sequestration offset standards
S. 1201 (Sanders) (CO_2 reduction from utilities)	Offsets are not included in the bill	Offsets are not included in the bill	Offsets are not included in the bill	Offsets are not included in the bill Directs EPA to implement emissions reduction program; directs Secretary of Agriculture, in coordination with EPA, to develop standards for biological sequestration

Table 2. (Continued)

GHG Reduction Program	Offset Types Allowed or Prohibited	Offset Quantity Limitation	Acceptable Locations of Offset Projects	System of Verifying Integrity of Offsets
S. 1554 (Collins) (CO_2 reduction from utilities)	Offsets are not included in the bill	Offsets are not included in the bill	Offsets are not included in the bill	Offsets are not included in the bill Directs EPA, in coordination with Department of Agriculture, to help develop procedures for verifying biological sequestration projects
S. 1766 (Bingaman)	Four specific project types shall have streamlined standards: (1) landfill methane; (2) animal waste or municipal wastewater methane; (3) sulfur hexafluoride reductions from transformers; and (4) coal mine methane; the President may add further types	Unlimited use of domestic offsets with identified standards; international offsets limited to 10% of a regulated entity's emissions target	Domestic and international	Directs the President to develop offset verification system; directs Secretary of Agriculture to establish agricultural sequestration standards
S. 2191 (Lieberman) Reported by the Senate Committee on Environment and Public Works May 20, 2008	Specifically allows certain agricultural and forestry-related offsets: agricultural land management practices; afforestation; reforestation; forest management; manure management; other offset types may be later allowed by EPA through regulations	Domestic offsets can satisfy 15% of allowance submission In addition, "international emission allowances obtained on a foreign GHG emissions trading market" can satisfy 15% of submission[c]	Domestic Indirect access to international offsets through purchase of international "emission allowances"[c]	Directs the EPA, in consultation with Secretary of Agriculture, to develop regulations to implement offset program; requires offset project developers to submit a petition to EPA and receive approval of project; offset projects must then be reviewed by an accredited

GHG Reduction Program	Offset Types Allowed or Prohibited	Offset Quantity Limitation	Acceptable Locations of Offset Projects	System of Verifying Integrity of Offsets
				third-party, who submits report to EPA for approval; reversal certifications must be submitted annually to EPA
S. 3036 (Boxer) H.R. 620 (Olver)	Same as S. 2191 Specifically allows for agricultural and conservation practices; reforestation; forest preservation; no limits on other types	Same as S. 2191 Up to 15% of allowances can come from domestic and/or international offsets; if offsets account for 15% of allowances, at least 1.5% must come from agricultural sequestration[d]	Same as S. 2191 Domestic or international	Same as S. 2191 Directs EPA—in coordination with the Secretaries of Commerce, Energy, and Agriculture—to develop verification methods and standards
H.R. 1590 (Waxman)	Offsets are not specifically addressed in the bill	Offsets are not specifically addressed in the bill	Offsets are not specifically addressed in the bill	Offsets are not specifically addressed in the bill; EPA is to ensure that allowances are accurately tracked, reported, and verified
H.R. 4226 (Gilchrest)	Specifically allows biological sequestration, which can include agricultural and conservation practices; reforestation; forest preservation; production of cellulosic biomass crops; and other methods determined by EPA; allows for use of other offset projects if approved and added to national registry; no restrictions on international offset types that are approved on case-by-case basis	Alternative compliance mechanisms, which can include domestic and international offsets, can account for up to 15% of allowance submission; if these alternatives account for 15% of allowances, at least 1.5% must come from registered sequestration in agricultural soils	Domestic or international	Directs EPA, in coordination with the Secretaries of Agriculture, Energy, and Commerce, to issue regulations that establish comprehensive measurement and verification methods Directs the EPA to develop program for reviewing international offset projects

Table 2. (Continued)

GHG Reduction Program	Offset Types Allowed or Prohibited	Offset Quantity Limitation	Acceptable Locations of Offset Projects	System of Verifying Integrity of Offsets
H.R. 6186 (Markey)	Domestic projects limited to: (1) agricultural activities (excluding those covered by performance standards, such as animal feeding operations); (2) afforestation and reforestation; (3) petroleum and natural gas system emissions; (4) coal mine emissions (excluding those covered by performance standards—i.e., mines emitting over 10,000 mtCO_2 per year) International offsets exclude (1) land use, land use change, and forestry activities; and (2) destruction of hydrofluorocarbons	Domestic offsets can meet up to 15% of a covered entity's allowance submissions; International emission allowances, international offsets, or some combination thereof can satisfy another 15% of an entity's allowance submission	Domestic or international International offsets can only come from nations that (1) have taken comparable emission reduction action; (2) emitted less than 0.5% of global GHG emissions in most recent year; or (3) identified by United Nations as among the least developed of developing countries	Directs EPA to implement offset program through regulatory process; EPA may require offset project developers to submit a petition to EPA and receive approval of project; offset projects must be reviewed by an accredited third-party, who submits report to EPA for approval
H.R. 6316 (Doggett)	Domestic offsets may include: (1) emissions from animal feeding operations; (2) afforestation and reforestation; (3) petroleum and natural gas system emissions; (4) coal mine emissions; and (5) other	Domestic offsets limited to 10% of allowance submission; agriculture projects limited further per NAS review; International forest allowances limited to 10% of allowance submission; Combined limitation of 25% of allowance	Domestic and international International forestry offsets only eligible from nations that meet specified conditions	Directs EPA, in coordination with the Secretary of Agriculture, to develop regulations to implement offset program, including procedures for project initiation and approval and third-party verification

GHG Reduction Program	Offset Types Allowed or Prohibited	Offset Quantity Limitation	Acceptable Locations of Offset Projects	System of Verifying Integrity of Offsets
	agricultural activities approved by a National Academy of Sciences (NAS) review International offsets include forest carbon activities: afforestation, reforestation, and avoided deforestation	submission from foreign allowances, international forest allowances, and domestic offsets		

[a] These projects would become ineligible if subsequent legislation required emissions reductions from these sectors (S. 317 only covers power plants).

[b] Offset standards similar to those required by the Regional Greenhouse Gas Initiative (RGGI), a partnership of 10 states from the Northeast and Mid-Atlantic regions. Unlike RGGI standards, S. 1168 does not require third-party verification for offset projects.

[c] The proposal does not define "international emission allowance." EPA is directed to develop regulations concerning their use.

[d] The legislation states that if an entity uses offsets to satisfy 15% of its allowances, "it shall satisfy up to 1.5 percent of its total allowance submission [with agricultural sequestration offsets].... " (Section 144(b)). This language is arguably unclear as to whether it limits ("up to") agricultural sequestration offsets to only 1.5% or requires that (at least) 1.5% of offsets come from agricultural sequestration activities.

Table 3. Comparison of Offset Treatment in GHG Emissions Reduction Initiatives in the U.S. States

GHG Reduction Program	Offset Types Allowed or Prohibited	Offset Quantity Limitation	Acceptable Locations of Offset Projects	System of Verifying Integrity of Offsets
Regional GHG Initiative (RGGI) (CO_2 reduction from utilities)	Five offset project types allowed:[a] 1) landfill methane reduction 2) Sulfur hexafluoride (SF6) reductions from industrial activity 3) Afforestation projects 4) Energy efficiency projects yielding reductions or avoidance of CO_2 from natural gas, oil or propane combustion 5) Avoided methane from manure management practices	RGGI was designed to require that 50% of emission reductions come from regulated sources; offsets can be used to achieve the remaining 50% of required reductions;[b] to apply this objective, offsets are limited to 3.3% of a source's emissions; the limit increases to 5% if the market price of an allowance exceeds \$7 (in 2005 dollars, adjusted annually); if price exceeds \$10 (in 2005 dollars, adjusted annually), the limit increases to 10%[b]	1) RGGI states; 2) non-RGGI states that have their own GHG reduction program or have signed a memorandum of understanding (MOU) with a RGGI state; 3) international projects (e.g., CDM certified emission credits) allowed if carbon price exceeds \$10	Standards Approach: each project must meet general standards and standards specific to the project type; each project must be certified by a third-party
California's Statewide Emission Program[c]	Not specified in statute; details deferred to California Air Resources Board (CARB); The Market Advisory Committee (MAC)[e] recommended starting with a small number of project types, such as those allowed under RGGI The CARB Scoping Plan[f] did not provide further details. Project types will be determined through a regulatory process.	Not specified in statute; details deferred to CARB The CARB Scoping Plan indicates that CARB will impose a quantity limitation on offset use; similar to the RGGI limitation, CARB will allow offsets and allowances from other programs to account for no more than 49% of emission reductionsg	Not specified in statute; details deferred to CARB Most MAC members rejected geographic limitations The Scoping Plan indicates that CARB supports the use of both international and domestic offset projects.	Not specified in statute; details deferred to CARB MAC recommended adopting a standards-based approach The Scoping Plan did not provide further detail; the offset system will be developed through a regulatory process

Note: Other states have recently enacted legislation to reduce GHG emissions, and a number of states have signed regional agreements that call for GHG reduction. However, these programs are relatively new, and the design details (in particular, offset treatment) have not yet been officially specified. Thus, they are not listed in the above table. For more information, on these programs, see CRS Report RL33812, *Climate Change: Action by States to Address Greenhouse Gas Emissions*, by Jonathan L. Ramseur.

[a] More projects may be added in the future.

[b] See RGGI Staff Working Group, *Analysis Supporting Offsets Limit Recommendation*, at http://www.rggi.org/documents.htm.

[c] The RGGI Memorandum of Understanding describes this increase in the use of offsets as a "safety-valve." Unlike a traditional safety-valve, the cap would be maintained because additional allowances cannot be purchased at a threshold price. The RGGI "safety-valve" would effectively allow regulated parties to meet the majority (at the 5% limit) or possibly all (at the 10% limit) of their reduction requirements through offsets: the RGGI cap is projected to require regulated sources to reduce their annual emissions by about 7% on average (based on RGGI Offsets Limits Analysis data at http://www.rggi.org/documents.htm.) The cost protection provided by RGGI's safety-valve will depend on the offset market. For example, if the supply of acceptable RGGI offsets cannot meet demands, the offset price may increase such that the safety-valve is negated. An assessment of offset supply and demand conducted by RGGI officials suggests that this outcome seems unlikely (*Evaluation of Offsets Supply and Potential Demand*, at http://www.rggi.org/documents.htm.)

[d] California Governor Schwarzenegger signed "The Global Warming Solutions Act" (AB32) into law September 27, 2006. AB32 creates a mandatory GHG emissions target: return to 1990 levels by 2020. The statute authorizes, but does not require, the use of market-based mechanisms. The California Air Resources Board (CARB) is responsible for crafting most of the logistical details, including offsets. For more information on AB32 see CRS Report RL33962, *Greenhouse Gas Reductions: California Action and the Regional Greenhouse Gas Initiative*, by Jonathan L. Ramseur.

[e] The MAC recommendations are included in the table for comparison purposes because the regulations are being developed. Per California Executive Order S-20-06, the Market Advisory Committee was formed to develop recommendations regarding design details for a market-based emissions reduction program. The Committee includes national and international experts with backgrounds in economics, environmental policy, regulatory affairs, and energy technologies. See Market Advisory Committee, 2007, *Recommendations for Designing a Greenhouse Gas Cap-and-Trade System for California*.

[f] In December 2008, the California Air Resources Board (CARB) approved a Scoping Plan, which was required by AB32 and outlines the activities that CARB will undertake to meet the GHG emission goals of AB32. The Scoping Plan may indicate the direction of future regulations, but each component of the plan will be developed through a regulatory rulemaking process, and thus is subject to

change. The Scoping Plan and related documents are available at http://www.arb.ca.gov/cc/scopingplan/document/scopingplan document.htm.

[g] Although this number may appear relatively high compared to federal cap-and-trade proposals, limiting offsets by percentage of reductions is different from limiting offsets as a percentage of allowance submissions (as many federal proposals would do). This 49% limitation is similar to recommendation made by the Western Climate Initiative. See Western Climate Initiative, *Design Recommendations for the WCI Regional Cap-and-Trade Program* (September 2008), at http://www.westernclimateinitiative.org/.

Table 4. Comparison of Offset Treatment in International Emissions Trading Programs

GHG Reduction Program	Offset Types Allowed or Prohibited	Offset Quantity Limitation	Acceptable Locations of Offset Projects	System of Verifying Integrity of Offsets
Kyoto Protocol	Clean Development Mechanism (CDM) projects: projects judged individually; wide range of types have been accepted; prohibits use of reductions generated from nuclear facilities;[a] land use, land use changes, and forestry (LULUCF) offset projects limited to reforestation and afforestation[b] Joint Implementation (JI) projects: may allow a broader array of project types than the CDM, and would include revegetation, forest management, cropland management and grazing land management; JI projects may be limited by a host country's emission control regulations	CDM and JI: "Supplementarity" constraint: offsets must be "supplemental to domestic action and that domestic action shall thus constitute a significant element of the effort made by each Party ... " (emphasis added);[c] but no specific quantity limitations CDM: Reforestation and afforestation projects limited to 1% of party's baseline emissions	CDM projects: developed nations finance projects in developing nations JI projects: developed nations finance projects in other developed nations; both nations must be parties to the Kyoto Protocol	Case-by-case approval process, including a test of "additionality"[d] CDM: each project must have letter of approval from both buyer and seller's governments; must be evaluated and approved by an Executive Board (EB);[e] independent third party (accredited by EB) determines the certified emissions reductions (CERs)

GHG Reduction Program	Offset Types Allowed or Prohibited	Offset Quantity Limitation	Acceptable Locations of Offset Projects	System of Verifying Integrity of Offsets
				JI: Track 1 - eligible host country may approve projects and assign emission reduction units (ERUs); Track 2 - Joint Implementation Supervisory Committee (JISC)[f] approves project and assigns ERUs
European Union's Emissions Trading System	Kyoto Protocol limitations listed above; Additional limitations: regulated sources cannot use offsets from land use, land use changes, and forestry (LULUCF) projects;[g] offsets from hydroelectric power projects must satisfy certain conditions	First phase (2005-2007): no limits for offsets from CDM, but no JI projects; Second phase (2008-2012) EU members set own limits for offsets from CDM or JI projects, but limit must fall within range set by the European Commission (EC):[h] at minimum EU states must allow regulated sources to use offsets to cover 10% of their allowances; at a maximum, offsets can cover up to 50% of the reductions required by cap (in some EU states this equates to 20% of allowances)[i]	CDM projects: developing nations JI projects: other EU nations Domestic offset projects (DOPs) not allowed[j]	CDM/JI offsets follow Kyoto Protocol verification process (described above)

[a] UNFCCC, 2001, Conference of the Parties, Sixth Session, Decision Five.

[b] UNFCCC, 2001, Conference of the Parties, Seventh Session ("Marrakesh Accords"), Decision 11. Afforestation involves planting trees on previously non-forested land; reforestation involves planting trees on formerly forested land.

[c] UNFCCC, 2006, *Conference of the Parties serving as the meeting of the Parties to the Kyoto Protocol on its first session, held at Montreal from 28 November to 10 December 2005*, Decision 2/CMP1.

[d] "Additionality" is a critical component of the environmental integrity of an offset. The concept refers to whether the offset project would have gone forward on its own merits (e.g., financial benefits) without the support of an offset market or the impetus to comply with a legal requirement. In other words, would the offset project have happened anyway? If the project would have occurred, the project is not additional, and should not qualify as an offset.

[e] The EB is composed of 10 members from parties to the Kyoto Protocol; the members' terms are limited.

[f] The JISC is composed of 10 members from parties to the Kyoto Protocol; the members' terms are limited.

[g] Although private parties subject to the ETS cap cannot purchase LULUCF offsets, EU governments can purchase eligible LULUCF offsets—i.e., from afforestation or reforestation projects—up to 1% of their state's base year (1990) emissions each year (See European Union Directive 2004/101/EC (October 27, 2004); Kyoto Protocol, Decision 17/CP.7 (November 2001)). The World Bank reported that global transactions of LULUCF offsets have only accounted for 6% of this allowable limit.

[h] European Commission Communication (COM/2006/725), November 29, 2006.

[i] If EU state governments purchase offsets (e.g., to sell as allowances for new sources), these offsets will reduce the percentage of offsets that can be used as allowances by affected sources within that state.

[j] This issue has received interest in recent months, and some EU members support including domestic offset projects. See European Climate Change Programme Working Group, 2007, Report of the First Meeting (March 8-9, 2007).

Another debate may focus on the possible effects of offsets in the developing world (assuming international offsets are allowed in a federal program). On one hand, many of the offset projects may offer significant benefits—more efficient energy infrastructure, improved air quality—to local communities. On the other hand, some maintain that if developed nations use all of the low-cost offsets in developing nations, the developing nations will face higher compliance costs *if and when* they establish GHG emission reduction requirements. Moreover, there is some concern that international offsets may serve as a disincentive for developing nations to enact laws or regulations limiting GHG emissions because they would lose funding from the offset market.

Whether to include international offsets in a federal program raises other considerations as well. The ability to use international offsets for compliance purposes would substantially expand emission reduction opportunities, compared to only allowing domestic offsets. The more emission mitigation opportunities available, the lower the carbon price. This highlights the debate over the balance between overall program costs and price signal for technological development.

If eligible in a U.S. program, international offsets from countries without binding reduction targets are likely to dominate in early decades because of their comparatively lower costs. Certain domestic economic sectors, primarily agriculture and forestry (if eligible as offsets), would benefit if international offsets are excluded. However, the inclusion of international offsets may benefit other U.S. economic sectors through the transfer of technology and services to support the projects. Moreover, as noted above, the more offset opportunities, the lower the overall costs of the cap-and-trade program.

End Notes

[1] In 1992, the United States ratified the United Nations Framework Convention on Climate Change (UNFCCC), which called on industrialized countries to initiate GHG reduction. The UNFCCC defines GHGs to include carbon dioxide (CO_2), methane (CH4), nitrous oxide (N2O), sulfur hexafluoride (SF6), hydrofluorocarbons (HFC), and perfluorocarbons (PFC).

[2] See CRS Report RL33826, *Climate Change: The Kyoto Protocol, Bali "Action Plan," and International Actions*, by Susan R. Fletcher and Larry Parker; CRS Report RL31931, *Climate Change: Federal Laws and Policies Related to Greenhouse Gas Reductions*, by Brent D. Yacobucci and Larry Parker; CRS Report RL33812, *Climate Change: Action by States to Address Greenhouse Gas Emissions*, by Jonathan L. Ramseur.

[3] See CRS Report RL30024, *U.S. Global Climate Change Policy: Evolving Views on Cost, Competitiveness, and Comprehensiveness*, by Larry Parker and John Blodgett.

[4] In this way, offsets would complement the more traditional emissions trading that can occur between two covered sources. For example, a covered source (e.g., power plant) can make reductions beyond its compliance obligations and then sell these reductions as credits to other covered sources. This type of transaction represents the "trade" component of a cap-and-trade program.

[5] An emissions cap might require only CO_2 emission reductions, but still allow CO2-e offsets from projects that involve non- CO_2 GHGs.

[6] Although Congress could address GHG emissions with alternative policies—e.g., by enacting a carbon tax or setting emission limits for each source type ("command-and-control)—the option to use offsets is generally discussed in the context of a cap-and-trade regime. Offsets could be a component of a carbon tax framework (e.g., as tax credits), but that discussion is beyond the scope of this report.

[7] For instance, if a covered source reduced its emissions beyond its compliance obligation, the source could sell the reductions as "credits" to other sources subject to the cap. This financial opportunity would create the incentive for sources to find and make reductions beyond their compliance obligations. These type of exchanges represent the foundation of the cap-and-trade system.

[8] For comparison purposes—e.g., estimating the quantity of offsets and potential offset benefits—this report generally assumes that emission sources and sequestration activities will either not be regulated in any fashion or they will qualify as offsets. However, there are alternative means of addressing emission sources and sinks that are often considered good candidates for offsets. See the Text Box on p. 8: Policy Alternatives to Offsets.

[9] For more information, see CRS Report RL34150, *Climate Change and the EU Emissions Trading Scheme (ETS): Kyoto and Beyond*, by Larry Parker.

[10] The credits are called "certified emission reductions" (CERs) or "emission reduction units" (ERUs), depending on whether they originate from the Clean Development Mechanism (CDM) or from Joint Implementation projects, respectively. The CDM is the Kyoto Protocol compliance mechanism, which has been used widely in the EU-ETS, that allows for developing nations to generate offsets and sell them to regulated sources in developed nations. The CDM was established by Article 12 of the Kyoto Protocol. For more information on the Kyoto Protocol's "flexible mechanisms," see http://unfccc.int /kyotojrotocol/mechanisms/items/1673.php.

[11] Although the credits are equivalent to allowances in environmental and economic terms, they are not interchangeable. For more on the EU ETS, see CRS Report RL33581, *Climate Change: The European Union's Emissions Trading System (EU-ETS)*, by Larry Parker.

[12] If the source was subject to an emissions cap, reductions beyond compliance obligations would be sold directly as emission credits.

[13] For more information on agricultural activities, see CRS Report RL33898, *Climate Change: The Role of the U.S. Agriculture Sector*, by Renée Johnson.

[14] See CRS Report RL34560, *Forest Carbon Markets: Potential and Drawbacks*, by Ross W. Gorte and Jonathan L. Ramseur.

[15] This comparison does not account for the externalities associated with fossil fuel combustion: air pollution, environmental degradation, health problems linked to emissions, etc.

[16] In addition, some may argue that nuclear energy could be considered a renewable energy. This debate is beyond the scope of this report.

[17] The digesters capture the methane, which can be used for energy purposes.

[18] GWP is an index of how much a GHG may contribute to global warming over a period of time, typically 100 years. GWPs are used to compare gases to carbon dioxide, which has a GWP

of 1. For example, methane's GWP is 25, and is thus 25 times more potent a GHG than CO_2. The GWPs listed in this report are from: Intergovernmental Panel on Climate Change, *Climate Change 2007: The Physical Science Basis* (2007), p. 212.

[19] Chlorodifluoromethane, referred to as HCFC-22.

[20] Of the CERs expected to be issued by 2012, the percentage drops to 22% (still the highest percentage by offset type). See the United Nations Environment Programme (UNEP), CDM Pipeline data, at http://cdmpipeline.org/index.htm.

[21] This calculus depends on the market price for offsets. See Michael Wara, *Measuring the Clean Development Mechanism's Performance and Potential*, Working Paper #56, Stanford Center for Environmental Science and Policy (2006).

[22] Footnote is not available.

[23] Footnote is not available.

[24] Footnote is not available.

[25] Intergovernmental Panel on Climate Change, *Climate Change 2007: Mitigation. Contribution of Working Group III to the Fourth Assessment Report* (2007), p. 526.

[26] Nonpoint source pollution is caused by rainfall or snowmelt moving over and through the ground. As the runoff moves, it picks up and carries away natural and human-made pollutants, finally depositing them into lakes, rivers, wetlands, coastal waters, and even underground sources of drinking water. See EPA's Nonpoint Source Pollution website, at http://www.epa.gov/owow/nps/qa.html.

[27] See CRS Report RL33800, *Water Quality Issues in the 110th Congress: Oversight and Implementation*, by Claudia Copeland.

[28] Intergovernmental Panel on Climate Change, *Climate Change 2007: Mitigation. Contribution of Working Group III to the Fourth Assessment Report* (2007), pp. 529-530.

[29] In fact, these activities are often categorized, particularly in international contexts, as land use, land use changes, and forestry (LULUCF) projects.

[30] For example, certain evergreen plantations (tree farms) generally have higher water use than the land they replace. Intergovernmental Panel on Climate Change, *Climate Change 2007: Mitigation. Contribution of Working Group III to the Fourth Assessment Report* (2007), p. 530.

[31] EPA, *Greenhouse Gas Mitigation Potential in U.S. Forestry and Agriculture* (2005), p. 8-8.

[32] As discussed above, the inclusion of international offsets would lower the emission allowance price, which would benefit regulated entities and ultimately consumers.

[33] http://www.planktos.com/. See Ken O. Buesseler, et al.," Ocean Iron Fertilization—Moving Forward in a Sea of Uncertainty," *Science* Vol. 319 (2008), 162.

[34] See, World Resources Institute, *The Greenhouse Gas Protocol for Project Accounting* (2005), at http://www.ghgprotocol.org.

[35] The emissions data may not be a direct measurement, but an estimate calculated by using related data, such as fuel consumption.

[36] See CRS Report RL34059, *The Carbon Cycle: Implications for Climate Change and Congress*, by Peter Folger.

[37] See CRS Report RL34560, *Forest Carbon Markets: Potential and Drawbacks*, by Ross W. Gorte and Jonathan L. Ramseur.

[38] Govindasamy Bala, et al., "Combined climate and carbon-cycle effects of large-scale deforestation," *Proceedings of the National Academy of Sciences*, Vol. 104 (2007): 6550-6555.

[39] See Anja Kollmuss, "Carbon Offsets 101," *World Watch* (2007).

[40] Frank Lecocq and Philippe Ambrosi, "The Clean Development Mechanism: History, Status, and Prospects," *Review of Environmental Economics and Policy* (Winter 2007), pp. 134-151.

[41] See Environmental Protection Agency (EPA), Office of Air and Radiation, *Tools of the Trade: A Guide To Designing and Operating a Cap and Trade Program For Pollution Control* (2003), Glossary.

[42] Nicholas Institute for Environmental Policy Solutions, *Harnessing Farms and Forests in the Low-Carbon Economy: How to Create, Measure, and Verify Greenhouse Gas Offsets*, Zach Wiley and Bill Chameides, eds. (2007), pp. 18-19.

[43] Similarly, forest sequestration projects could shift demand to substitute products (e.g., steel or aluminum studs to replace wood studs in homebuilding) whose production requires more energy, and thus releases more carbon. See CRS Report RL31432, *Carbon Sequestration in Forests*, by Ross W. Gorte.

[44] Testimony of David Hawkins, Climate Center, Natural Resources Defense Council, before the Senate Committee on Environment and Public Works, November 13, 2007, at http://docs.nrdc.org/globalwarming/glo_07111301A.pdf.

[45] These are the costs assessed in the following study: Camille Antinori and Jayant Sathaye, *Assessing Transaction Costs of Project-Based Greenhouse Gas Emissions Trading* (2007), Ernest Orlando Lawrence Berkeley Laboratory.

[46] Ibid.

[47] See e.g., David M. Driesen, 1998, "Free Lunch or Cheap Fix?: The Emissions Trading Idea and the Climate Change Convention," *Boston College Environmental Affairs Law Review* 26:1-87; see also Emily Richman, 2003, "Emissions Trading and the Development Critique: Exposing the Threat to Developing Countries," *New York University School of Law Journal of International Law and Politics* 36:133-176.

[48] See e.g., Ross Gelbspan, "Toward A Global Energy Transition," *Foreign Policy In Focus* (2004).

[49] This attribute also creates critical challenges for policymakers. For instance, if one nation invests in emission reductions, any resulting benefits (e.g., decreased atmospheric GHG concentration) would be shared by all nations, including those that continue to increase their emissions. This dynamic has led some to refer to climate change as the "ultimate global commons pollution problem" because it discourages unilateral emission reduction. See Henry Lee, 2001, "U.S. Climate Policy: Factors and Constraints," in *Climate Change: Science, Strategies, & Solutions* (Eileen Clausen, editor).

[50] For a discussion of the potential supply of offsets in a cap-and-trade program, see CRS Report RL34705, *Potential Offset Supply in a Cap-and-Trade Program*, by Jonathan L. Ramseur.

[51] National Commission on Energy Policy, 2007, *Energy Policy Recommendations to the President and the 110th Congress*.

INDEX

#

A

B

C

D

E

I

J

K

L

M

S

T

U